Research and Growth in Agriculture Production : A Case Study

Research and Growth in Agriculture Production : A Case Study

M.S. Rana

CYBER TECH PUBLICATIONS
4264/3, Ansari Road, Daryaganj, New Delhi-110002 (India)
Ph.: 011-23244078, 011-43559448 Fax: 011-23280028
E-Mail: cyberpublicationsdelhi@yahoo.com
Website: www.cybertechpublications.com

RESEARCH AND GROWTH IN AGRICULTURE PRODUCTION : A CASE STUDY

M.S. Rana

First Edition 2010

Published by : Asian Offset Printers

G.S. Rawat for **Cyber Tech Publications**
4264/3, Ansari Road, Daryaganj, New Delhi-110002 (India)
Ph.: 011-23244078, 011-43559448 Fax: 011-23280028
E-Mail: cyberpublicationsdelhi@yahoo.com
Website: www.cybertechpublications.com

Preface

India's investments in agricultural research, extension, and irrigation have made it one of the largest publicly funded systems in the world. But some policymakers who perceive that the benefits to research may be declining are advocating a cut back on public spending on research. This research report, which examines the effects of research and development on productivity in India, finds that India is still benefiting from these investments.

The main sources of agricultural productivity growth in India during 1956–87 were public agricultural research and extension; expansion of irrigated area and rural infrastructure and improvement in human capital were also important contributors.

The report also shows that the public benefits from private research can be sub stantial, indicating that private firms capture only part of the real value of improved inputs through higher prices. Private agricultural research accounted for more than 10 per cent of growth of total facto productivity (TFP) during 1956–87, and in 1966–75, when India was more open to foreign technology, private research contributed 22 per cent of productivity gowth.

Industrial policy and technology policy, including intellectual property rights policy, will require careful evaluation and reform in order to encourage private investment in agriculture. Even so, Pray and Rosegrant argue that barriers to technology transfer should be removed in order to stimulate technology transfer and growth. Nevertheless, public investment in agricultural research will likely retain its primary role.

Contrary to concerns that growth in TFP has decreased over time, the report finds that during 1977–87, the period when the results in regions that adopted high-yielding varieties early on could be expected to taper off, TFP growth was 50 per cent higher than before the Green Revolution and 17 per cent higher than in the early years of the Green Revolution, indicating that gains are far from over.

The rates of return to public agricultural research are high, and it appears that the government is under investing in agricultural research. Expanding public investment in research and extension would lead to even greater gains."

— *Author*

CONTENTS

Preface

1. Introduction 1
2. Land Degradation due to Hydro-salinity in Semi-arid Regions 28
3. Irrigation Performance of Rice-Based Bhadra Project in India 37
4. Assessing biologically Degraded Soil 69
5. Research on Mobile Mapping System and its Application in Precision Agriculture 75
6. Rainfed Farming Development 83
7. Trends in Recent Agricultural Developments 93
8. Trends in Horticulture Growth 143
9. Agricultural Engineering 159
10. Agricultural Education and Research 183
11. Agricultural Marketing and Agri-business 197
12. Researching the Drylands 212
13. Policies Toward GM Crops in India 241

Bibliography 277

Index 279

1

INTRODUCTION

Agriculture is the life blood of the National economy as more than 70% of the population depends upon agriculture for their livelihood.

AGRICULTURAL PERFORMANCE

Desired emphasis has been given to agricultural sector since the inception of Five-Year Plan periods due to the agrarian driven economy; even today agricultural sector fuels GDP growth by contributing over 15 per cent (2002-03). Agricultural growth enhanced overall economic growth of the state during 1980s but its contribution in real terms to overall economic growth has declined during 1990s. Performance of Tamil Nadu's Agricultural Sector has been impressive since 1960's when early improved crop varieties were introduced. After the introduction of improved varieties, a phenomenal breakthrough in productivity of crops was achieved resulting in higher production of most of the crops.

Tamil Nadu has done extremely well in irrigated agriculture particularly in rice, sugarcane and groundnut, the major crops of the state. Foodgrains production registered a growth rate of 2.31 per cent beginning green revolution period until 1980s, thereafter it rose further to 3.71 until 1990s. This means, the foodgrains production was much ahead of population growth during the above period. Among the cereals, much jump in production of maize was noticed in 1990s as compared to 1980s since demand for poultry and other livestock products pushed the demand for maize as livestock and poultry feed.

A sharp decline in rice area from the peak of 28 lakh hectares to less than 20 lakh hectares is mainly caused by Cauvery water dispute and it is the main reason for slowing down in growth of foodgrains production. Lesser profitability of sorghum, cumbu and ragi due to poor demand for these grains, as the underlying factor, caused the significant decline in the area of these crops. However, impressive performance in yield rates of foodgrains has more than compensated the fall in area. As a result, production of foodgrains increased over the period. Technology, expansion in irrigation and market development paved the way for perceptible growth in yields of rice, sugarcane and groundnut. In pulses also, the productivity growth is high in Tamil Nadu. At present Tamil Nadu is competing with Punjab in rice productivity where there is perennial source of irrigation. The state has made remarkable progress in achieving higher productivity of sugarcane and groundnut on par with the productivity levels at all India level. The productivity of these crops has been stabilized in the irrigated areas. To achieve the targeted growth rate in agriculture, new initiatives like the private sector participation in precision farming and contract farming have been taken up. Agricultural production is also linked to both domestic and external markets through establishment of agricultural export zones for floricultural and horticultural commodities.

CONSTRAINTS

Marginalization of land holding, high variability in rainfall distribution, inadequate capital formation by the public sector, declining public investment on agriculture, declining net area sown, over - exploitation of ground water and inadequate storage and post harvest facilities affect the agricultural performance in the state. The state supports seven per cent of the country's population but it has only four per cent of the land area and three per cent water resources of the country. Of the total gross cropped area, only 50 percent of the area is irrigated in Tamil Nadu. Similarly, of the total area under food grains, only 60 percent of the area is irrigated. Nearly, 52 per cent of area is under dry farming conditions in Tamil Nadu apart from stable cropping intensity which is hovering around 120 per cent over the period. In spite of the above constraints, the State has made a tremendous performance in the production of crops, which is attributed mainly to the productivity increase.

CHALLENGES

There is a need for improving the agricultural growth to meet the challenges like rising food demand, increase in industrial raw material requirements and providing many gains in employment in agriculture through galvanizing potentials of agriculture. Tamil Nadu did well in irrigated agriculture, while rainfed agriculture has not been given adequate attention in terms of infrastructure development, technology and extension delivery system resulting in low agricultural growth and high level of poverty. Coverage of High Yielding Varieties was higher during early phases of green revolution and increased tremendously in the irrigated environment compared to rainfed environment. There are deficits in supply of certain agriculture commodities like pulses, oilseeds and cotton. Increase in pulses production is marginal and the state has to depend on imports to meet the demand. Cotton production in the state declined at the rate of 2.38 per cent per annum during the last two decades mainly due to decline in area by 2.57 per cent. Productivity growth of cotton is marginal (0.20 per cent) and average productivity of cotton in the state is lesser than the national average particularly due to large scale cultivation under rainfed condition. Major oilseeds such as groundnut and gingely exhibited declining tendency in terms of area and production. The area under fallow lands has increased over the years due to various reasons. The current fallows are on the increase from 12.02 lakh hectare in 1970s (9.2 per cent) to 15.03 lakh hectare in 2002-03 (11.6 per cent) ; area sown more than once tended to decrease from 13.21 lakh hectare in 1970s to 6.01 lakh hectare in 2002-03. The gross cropped area which reached the maximum in 1970s with 74.56 lakh hectare sharply fell to 51.91 lakh hectare in 2002-03.

OPPORTUNITIES

Improving the efficiency of water use, increasing the effectiveness of public expenditure and agricultural extension and spurring the development of agricultural markets are the key factors for achieving targeted growth in agriculture. Improved irrigation management practices and irrigation technologies (such as drip and sprinkler irrigation) and new investment in canals and water storages (coupled with improved operation and maintenance) will be crucial for improving the water use efficiency in the state. Use of sprinkler and drip technology has been promoted in the state to a certain extent. High capital cost of these technologies currently poses a constraint to widespread adoption by marginal and small farmers. Therefore, greater attention to development of more affordable technologies is warranted or a suitable system of targeted subsidies should be developed to scale-up the use of sprinkler and drip systems.

Contract farming is emerging in many of the crops like maize, cotton, medicinal plants, etc as an alternative system of crop production. Diversification into less water intensive high value products including fruits, vegetables, spices and condiments is one of the most promising strategies for increasing agricultural growth in Tamil Nadu. Horticulture is accorded a very significant sector in the emerging economic scenario due to existence of tremendous potential for export of exotic horticultural products especially fruits, flowers and vegetables. Another important factor in favor of development of the plantation and horticulture sector is the propensity of increasing the income levels of the farmers in rainfed areas by taking up dry land horticultural crops in dry lands as an alternate land use system as well as on cultivable wastes and other fallow lands. Such diversification could induce private investment in the processing for many of the higher value products in agricultural sector, which is likely to generate new rural non farm employment opportunities and contribute to higher rural incomes. Rich resource endowments offer vast vistas of opportunities for setting up of processing industries to produce value added products. Target research and Public Private Partnership (PPP) is necessary for attaining the momentum of growth in agriculture which has been targeted 4 per cent in the tenth plan period.

GIS BASED DIAGNOSTIC ANALYSIS OF IRRIGATION SYSTEM PERFORMANCE ASSESSMENT

Satellite remote sensing (SRS) and geographic information (GIS) techniques for improved water management in canal irrigation schemes. The Bhadra command area was brought under NWMP(National water management project) to improve agricultural productivity and farm income through a predictable, equitable and reliable irrigation service. Satellite remote sensing technique has been applied to historic and 1995 Rabi season data by National Remote Sensing Agency to generate primary data on irrigated area, cropping pattern and crop yield at disaggregated level and to access the improvement in agricultural productivity and water management after MWMP implementation. The GIS technique helps in integration of satellite and ground information to evaluate the system performance and to diagnose the inequality in the performance to aid in improving the water management.

The primary objective is to diagnose the factors for poor performance of selected distributaries using satellite data and s ground data collected by specially designed sample survey and to improve performance by prescribing corrective measures. The distributaries 9A, 12 and 15 of Malabennur division of Bhadra project in Karnataka state covering about 2350 ha area, is considered for this study. The performance of 12th distributory is good whereas the other two were poor performing. Hence to find the causative factors for poor performance were studied in comparison with a good performing area.

The satellite inputs (crop map and condition map) geometrically corrected wit respect to topographic maps are generated with ARIES and EASI/PACE software and transported to PAMAP environment. The revenue survey map in 1:8000 scale is photographically reduced to 1:16000 scale, digitized, edited and corrected with respect to top map and satellite imagery. Conducting sample survey designed for this purpose and compiled in Dbase collects the ground data for the analysis of identified problem distributaries. The system performance of the problem distributaries are characterized based on satellite reported acreage and yield gaps and analyzed with reference to ground sample survey data. The diagnostic analysis of the problem distributaries is carried out on the aspects of irrigation, agriculture and socio-economic. A questionnaire to this effect was designed and the information collected from the ground was organized and analyzed in the GIS. The causative factors are identified and prioritized for corrective measures to ensure better performance within the framework of NWMP. The analysis is performed by comparing the good with the bad performing distributaries.

The physical and socio-economic data were interpolated with different considerations and methods and analyzed. A Conjunctive analysis of remote sensing- derived, irrigation, agricultural and socio-economic data was attempted. Various causative factors pertaining to all these three activities were identified, ranging from poor physical condition, improper fertilizer consumption to poor interaction between the officials and the farmers. Here GIS served as a platform to analyze the effect of each component of the system based on the results.

NEED FOR IRRIGATION WATER MANAGEMENT

Irrigation water use is by far the largest use of water by mankind worldwide. The ever-increasing water demand compared with the depleting water resources warrants refined water use practices in irrigated agriculture to attain improved socioeconomic benefits. In the past years the improvement in the irrigation system concentrated in the hardware component namely the physical aspects like structures but limited on the software part namely water usage for agricultural purpose.

Water is not a free commodity. With increasing standards of living and fast growing population, the available water resources may not be able to meet various demands of mankind. It becomes necessary to put the available resource more effectively for more benefits. It is unto the managers of water resources to devise ways and means of optimally using resources to meet the ever-increasing demand. The aim of efficient irrigation water management or precisely, maximum yield with available water. A good management, proper and timely application of water may result in better yield and reduction in drainage problems.

Fig 1: Match between water release and agriculture operations

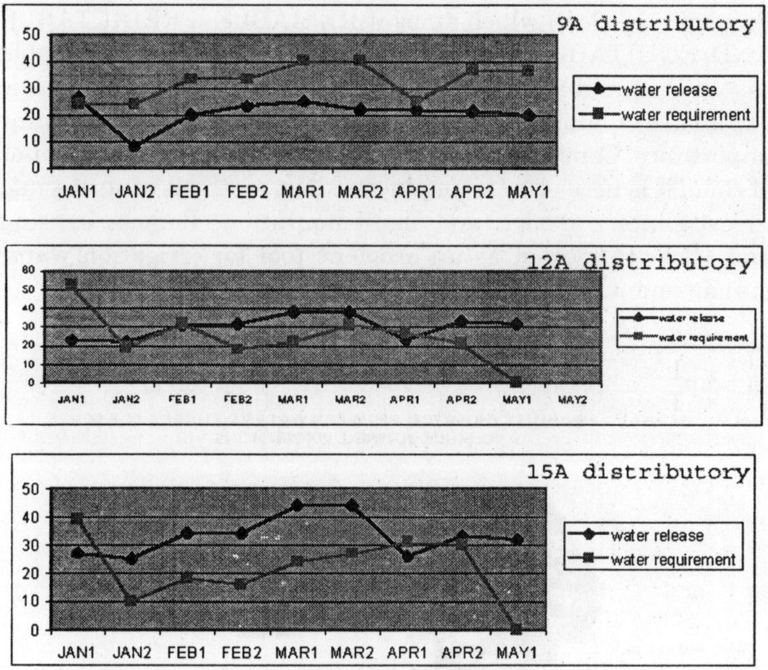

Actually the irrigation system performance is characterized by the effectiveness of the irrigation throughout the season in a timely pattern averting the losses thereby improving the economic gains (Martin Hvidit, 1997). If the control over irrigation water and its distribution in time and space according to the cropping pattern can be exercised, the system output in terms of production and economic gains will be realized. On the other hand, the water availability in time and space will command the agricultural practices such as cropping pattern, calendar, etc. Such a synchronization of cultivation and irrigation is possible only when the water control in terms of adequacy, timeliness and uniformity is practiced in a command. For this purpose the irrigation system operation should be so devised to create a potential for high performance irrigation resulting in an application system that farmer can use to irrigate his field uniformly in right time, with right amount (Skewes M.A, 1998). With this viewpoint, government of India with assistance from World Bank launched the "National Water Management Program (NWMP) which aims at RELIABLE, PREDICTABLE AND EQUITABLE supply of water. The timely and reliable assessment and monitoring of water resources and systematic exploration and developing new ones is of paramount importance. Considering the stupendous task and the constraint of time, it is necessary to employ modern methods of surveying, investigations, design and implementation. Remote sensing and GIS are viewed as an efficient tool for irrigation water management.

Fig. 2: Analysis on the cultivation – irrigation practices

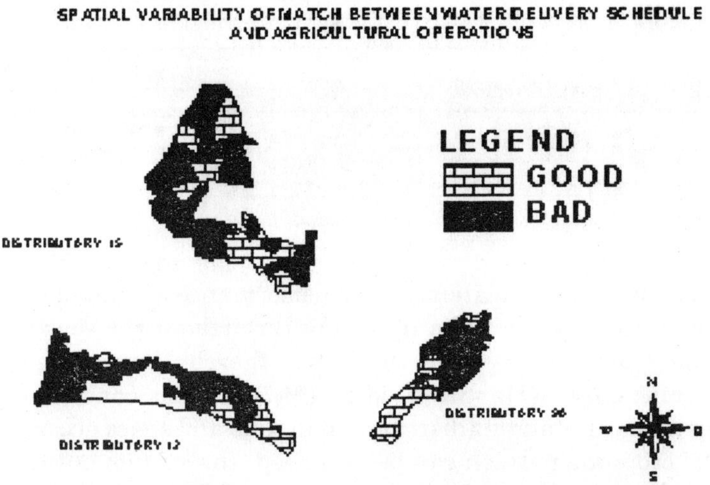

Remote sensing techniques are useful for Irrigation water management in the following areas (NRSA 1993)

- Assessment of water availability in reservoir for optimal management of water to meet the irrigation demand.
- Identifying, inventorying and assessment of irrigated crops.

- Determination of irrigation water demand over space and time
- Distinguishing lands irrigated by surface water bodies.
- Estimating crop yield
- Water logging and salinity problems in irrigated land
- Irrigation system performance evaluation

Fig. 3: Analysis on the crop performance

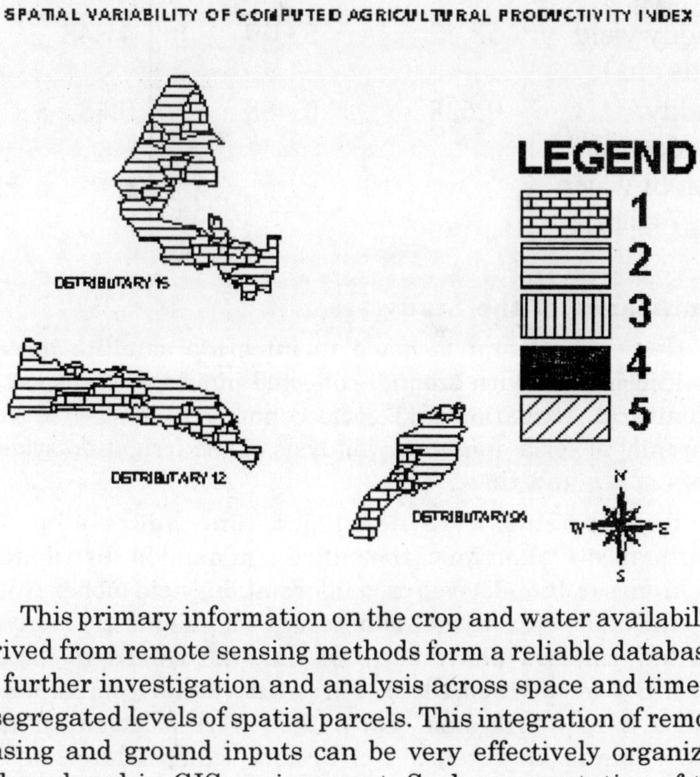

This primary information on the crop and water availability derived from remote sensing methods form a reliable databases for further investigation and analysis across space and time at desegregated levels of spatial parcels. This integration of remote sensing and ground inputs can be very effectively organized and analyzed in GIS environment. Such augmentation of the basic information of the system and the expert knowledge results in a system that addresses the present scenario of the system and future development in the system (Wolfe D.S , 1997). One of the advantages of GIS is that it gives a clear picture of effect of any action plan employed in the system through "if-then " simulation.

Table 1. The performance report of three distributaries

Performance indicators	9A distributary	12 distributary	15 distributary
Irrigation Intensity	1.026	0.716	0.515
Paddy area(ha.)	574	468	535
Paddy yield (qtls / ha)	58.42	53.49	41.48
Paddy production / unit of water (kg / cu.m)	0.628	0.468	0.483

Significance of the Study

Here an attempt is made to integrate satellite-derived crop information with ground- collected inputs with respect to agricultural, irrigation and socioeconomic practices, in GIS environment for a diagnostic analysis of the irrigation system across space and time.

Two systems were developed; one addressing the performance evaluation of the entire command at distributary level from satellite-derived crop information, yield model, water use efficiency, etc.; second addressing the diagnostic analysis of a few selected distributors to find the causative factors (Chari S.T et al, 1995,1996). Three distributor commands of Bhadra Irrigation system, Karnataka were selected for this study. This is an extension of system which addresses the performance evaluation of Irrigation command and compares the good and poor performance in distributaries reported by the first system. An analysis is carried out to analyze these distributories with respect to report practices add the performance to find out the causative factors for poor performance. This is to basically establish GIS as a tool aid analysis irrigation system at different levels and to help the time- manned location-wise decision-making process for enhancing the efficiency of the system.

Fig. 4: Analysis on fertilizer application

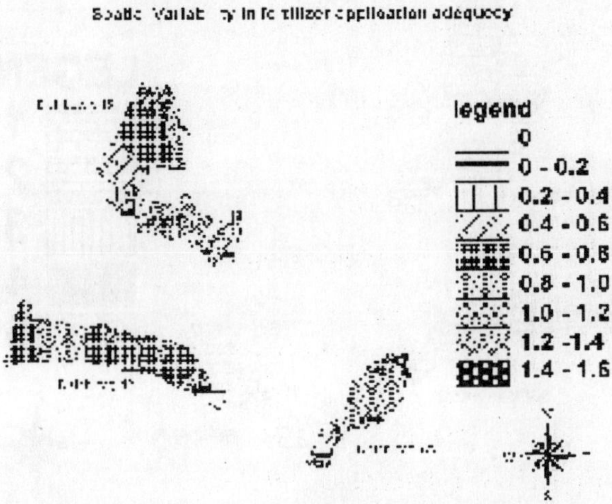

STUDY AREA

The Bhadra Dam is located across Bhadra river near Lakkavalli Village , Chikmagalur district of Karnataka state, at an elevation of 610m above MSL. Bhadra project comprises of a storage reservoir with a capacity of 2025 m3 a left bank canal and a Right Bank canal. The three distributories for the diagnostic analysis having no. 9A, 12 and 15 belonging to Malebennur branch canal of the right bank canal. The Malebennur branch canal is 48km in length and a full supply depth of 2.44m. The full supply discharge is 20.10 cusec and the irrigated area is 23,710 ha. The 9A distributory is in a very good condition with about 80% cement concrete lining. The minors are lined to some portion. The 12 and 15 distributaries are partially lined with about 10 m portion being lined on both sides of the PO. The 15 distributary is in a very poor condition with the entire portion of the channel being weeded and with lot of physical damages to the system. Farmers at the head reaches in all the three distributaries are going only for paddy and those at 15 distributary for crops like sugarcane and other semi-dry crops.

Fig. 5: Analysis on farmers involvement

SPATIAL VARIABILITY OF COMPUTED AGRICULTURAL PRODUCTIVITY INDEX

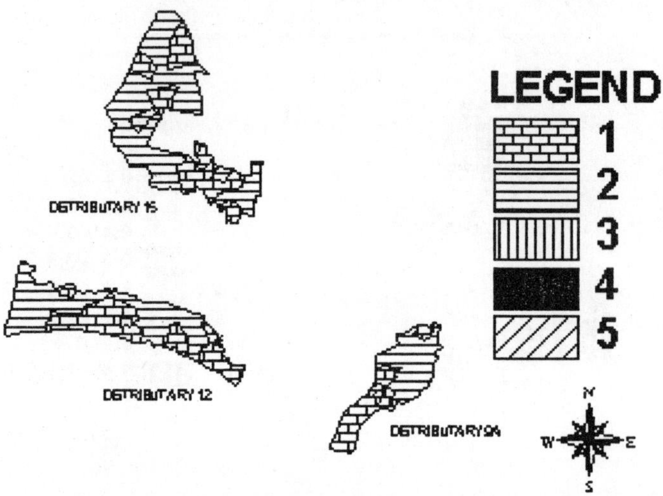

Probable Reasons for the Contrasting Performances of the Distribution

Three distributaries selected for the study and their performance indicators are listed below in table 1.

The following are the 15 probable reasons identified for the contrasting performances of the different distributaries under almost similar climatic and agro_ecological conditions:

- Realistic operational plan
- More water availability / unit area
- Better enforcement of the operational plan
- Better canal condition
- Less water stresses during the critical stages
- better soil fertility conditions
- Better water management
- Better application of fertilizers as per requirements
- Agricultural calendar in accordance with the water availability
- Better agricultural extension works

- Equitable distribution of water
- Pickup's / alternate water sources
- Extent of NWMP works
- Better credit facilities
- Farmers NWMP involvement with regard to:
- Awareness
- Involvement
- Response

DESIGN AND ORGANIZATION OF THE DETAILED QUESTIONNAIRE

To obtain all the important information for the diagnostic analysis, the questionnaire based on the experience gained from the reconnaissance survey is designed for the irrigation officials, agricultural authorities and the farmers soak s to obtain information from the relevant sources.

Table - 2: Classes for farmers involvement index and their relative weightages

Farmers awareness	Aware (1)	Not aware (1)
Farmers response to Irrigation	Schedule Good (1)	Not Good (2)
Status of Agricultural extension works	Good (1)	Not Good (2)
Farmers views in Irrigation department	Good (1)	Not Good (2)
Farmers views on Agricultural department	Good (1)	Not Good (2)

The questionnaire to the irrigation authorities covers questions related to irrigation water management, canal conditions, interaction with the farmers and the agricultural officials. The questionnaire prepared to the agricultural authorities covers questions related to the crops to be grown in the study area, their interaction with the farmers and the irrigation officials, type of agricultural extension and other miscellaneous information regarding the soil fertility conditions in the study area. The questionnaire executed to The farmers covers apart from other information, information related to their identification, the details regarding the inputs and the production in the current season, their involvement in the various NWMP activities, their views about the irrigation and the agricultural authorities, and their knowledge about the various agricultural aspects.

Digital Map Base Preparation

The digital map base preparation is the first step towards the presentation of a GIS module for the irrigation water management. The map base is prepared by scanning the topo maps and the available revenue survey maps of the three distributaries and are reduced to a scale of 1:16000 to 1:8000 and are digitized edited and corrected with respect to topo maps. From the base of revenue levels, the base maps of reach and distributary level are prepared. The three map bases are stored in 3 levels of GIS. The revenue survey numbers were given a reference numbers in the GIS map base so that the data transfer can be done. A total of 878 polygons was formed in the base map and was given sequential numbering. The distributory map base consists of 3 polygons, each representing one distributary. Each polygon was given a reference for data transfer. Similarly, the reach map base consists of 9 polygons and was also sequentially numbered.

SAMPLING PROCEDURE FOR THE SELECTION OF THE FARMERS

The questionnaire designed is to be executed for the farmers in the 3 distributaries to get the attribute data. However, the total number of farmers in the 3 distributaries sum up to 1000 forcing the sampling to ease the complexity involved in the collection of data. The sampling done for the present study is a sort of stratified random sampling based on the yield variability such that the reaches having high variability in yield have more samples. Once the number of farmers in a particular reach were selected, the further selection was totally based on random sampling so as to see that the farmers selected were having their evenly located within a reach.

Detailed Field Survey

A total of 100 farmers were selected for the detailed field survey. Apart from this, the questionnaires to the agricultural and the irrigation authorities were also executed to the concerned officials. The questionnaire to the farmers was executed by the team comprising of 4 scientists, two having water resources background and other 2 with agricultural background. During the interview, questions on farmer's personal views and wishes were noted and the farmer's personal experiences with the dept. people were also noted. The collection of data from the irrigation department was done by distributing the questionnaires containing field conditions and field operations, to various officials at different Levels. Office data in the form of discharge data, structural condition records and other relevant data were also collected. The questionnaire to the officials of the agricultural dept. was administered through personal interview to the asst. agricultural director. The data such as soil fertility tests records for the areas covering the 3 distributaries was also collected.

Interpolation of Physical Resources Data

As a first step, the ground reported yield was interpolated using the weighted average interpolation technique to generate a yield map. This ground report yield map is cross checked with the satellite-driven yield to:

- Prove the validity of yield model and
- Estimate the error of interpolation

Ground report travel time of flow was also interpolated to get a general picture of the canal condition, the moisture stress to get the general picture of the moisture stress,the fertilizer application adequacy (fig. 3) to find out the spatial variability of fertilizer application which gives an idea about the effect on yield and the involvement of farmers in various extension works.

Cluster analysis for socio-economic data: Socio-economic factors such as farmer's awareness, response and involvement in various NWMP works and views on the various govt. agencies are spatialised by performing a cluster analysis using theissen polygon method. Using third method, the following maps were generated

- Farmers awareness to the various program's under NWMP
- Match between water delivery schedule and agricultural operation
- Farmers involvement in irrigation scheduling
- Status of agricultural extension works
- Farmers views on the responsiveness on the dept. of irrigation

These clustering enable the zonation of the study area into zones of farmer's level of involvement which was studied for farmers awareness and responses to the various NWMP program's. To obtain all the important information for the diagnostic analysis, the questionnaire based on the experience gained from the reconnaissance survey is designed for the irrigation officials, agricultural authorities and the farmers soak to obtain information from the relevant sources. Then systems such as digital base map preparation, sampling procedure for the selection of the farmers, detailed field survey, interpolation of physical resources data, cluster analysis for socio-economic data were performed.

ANALYSIS ON THE PRIMARY AND DERIVED DATA

The data analysis was performed in two different environments. The primary or the first level analysis was performed in dBase III plus where the typical and most important trends for the performance of three distributaries were analyzed. The second level was in the GIS environment to cluster and group the inputs. The following analysis were performed

- Water availability per unit area: Here the water at the take off point was taken as the conveyance and application losses were not measured. This analysis showed that water available per unit area for 15th distributory was less compared to the other areas. The further analysis was done now based on this as one of the main constraints.

- Principal Component Analysis: This was done to generate the indices showing the major causative component. Two types of indices namely Agricultural Productivity index (API), to study the effect of inputs on productivity and Farmers Involvement Index (FII) , to study the farmers participation In the irrigation activities.
- The API was derived based on the inputs related to water, seeds, etc. The Correlation of these factors with yield is taken weightage factor.

API = W (w1) + F (w2) + s (w3) + P (w4)

API = Agricultural Productivity Index

W = Water Availability (moisture stress);

F = Fertilizer application adequacy

S = seeds quality;

P = Pesticides usage

w1 to w4 = respective weightages

A closer look at the API map (fig. 3) in conjunction with the yield map reveals that the inputs used by in the 15th distributary were not good resulting in the low yield.

- The FII was calculated by taking into consideration the awareness, response and involvement in the NWMP works. The weightages for various factors applying Boolean logic.

The above table 2 shows the classes for the various factors and the weightages for various factors. The importance of this index, as shown by fig. 5 is related to the review of the farmers status and responsiveness Particularly with yield information, this would indicate the future coordination and education that has to be taken up for better results.

- Match between the agricultural calendar and the canal operation plans were analyzed through the season for the three distributaries. From the graph below it can be observed that the two major and basic component of the system do not match in 12th and 15th distributory. Spatial mapping for this also done to examine the activities in the command.

- Water stress during the critical stages were also studied from the satellite and ground information. The table below gives an important conclusion that water stress in the 15th distributary was high which might have serious implications on the overall performance. The water stress in the 12th distributary were also high during flowering and the grain formation which could have had an impact on the yield.
- Spatial variability of the yield at different reaches of the distributaries were studied to locate the low yielding pockets and further analysis with respect to gap in structural and logistical components.
- The fertilizer application adequacy as informed by the field was mapped to understand the agricultural activities (fig 4) The fertilizer application was very varied from 20% to 140% suggesting the extent and the type of extension works.
- When analyzed in conjunction with yield details, the poor performance of the middle reach of the 15th distributary could be attributed to poor application.
- The agricultural operation reported from eh field were analyzed with respect to the canal delivery schedule and following observations made
- The match (fig 1) between water delivery scheduling and the agricultural operations was bad in the middle and tail reaches of 15th distributary.
- This was good in 12th distributary, could be attributed to the better yield figures.

Discussions on Result

The timely and reliable assessment and monitoring of water resources and systematic exploration for development are of paramount importance. Considering the stupendous task and constraint of time in developing the ultimate irrigation potential, it is necessary to use the modern methods of surveying and analysis tools. Remote sensing and GIS with their capability of data collection and analysis are now viewed as efficient and effective tools for irrigation water management. The capability of GIS to analyze the information across space and time would help in managing such dynamic systems as irrigation systems. The study shows the efficacy of this tool to analyze the information on irrigation system in various domains in an integrated manner to understand the system and upgrade the activities. Much of the research has been done on augmenting the resource information and analyzing them for better management . It is also required to improve the methodologies to realize the field –important information in Gis environment.

In this study , an attempt has been made to analyze the irrigation system from the viewpoints namely, irrigation, agriculture and socio-economics. Any system needs to account for the human or socio-economic component to realize the fruit of the effort. The capability of SRS and GIS to address the complex spatial problems and to assist in spatial decision making process qualifies it to handle water resources problem. Water management will be successful only if the human part of the system is considered. This study shows that the operation plan and its adherence had a greater impact on the overall performance of the system. The analysis on the match between the operation plan and the agricultural operations (fig 1) illustrates the efficacy of the Gis to handle the data across space and time. The decision-making and statistical approaches used for formulating the indices that would have a positive relation to the performance within the system is one example of integration. The study also showed that various socio-economic patterns can be derived and studied for management purpose. One important observation is that the farmers views on the Government Agencies were better in the areas of good yield. This shows the direction to be taken by the governmental agencies to bring about higher and sustainable productivity.

Scope

The sampling procedure can be improved after studying various socio-economic patterns of the area under study to have faithful representation of the population. The stratification for sampling can be done on the basis of more than one criteria. More time and effort can be spent for the field survey and interview with farmers, as this is the basic input for the system.

More time would have educated the farmers with respect to the importance of their response. More queries can be added to the questionnaire after consulting the experts in this particular field of study.

The spatial statistical techniques applied to convert the farmers response to queries to surface pattern itself need to be studied in detail. This holds paramount importance in deriving the indices for diagnostic analysis. Also the indices derived have to refine with the help of socio-economic and statisticians. This particular area of spatial statistics has greater scope in the years to come.

PERFORMANCE EVALUATION OF AN IRRIGATION PROJECT USING SATELLITE REMOTE SENSING GIS & GPS

India is amongst the largest irrigator countries in the world today. There is, however, increasing concern about some of the irrigation potential created not brought into the functional system, low operating efficiency, less crop productivity etc. System performance monitoring, evaluation and diagnostic analysis are keys to appreciate the improvement or inefficiency in our irrigation projects.

Irrigated lands' baseline inventory in spatial and time domains using spatial information technologies (satellite remote sensing, digital image processing, GIS and GPS) provides an array of performance evaluation matrices to address this issue. Case study in Nagarjunasagar irrigation project in Andhra Pradesh, India is cited as a realization of this modern information technology tool.

A large number of irrigation projects have been commissioned in India in the post-independence era for improving food production and economic development. These irrigated croplands, though limited to about 30 percent of the gross cropped area in India, contribute more than 50 percent of the total agricultural production.

This means, scientific management of irrigation water and irrigated croplands is the only way to make our agriculture productive and competitive. It is realized, however, that the irrigation potential created is not being fully utilized and gap exists between the potential created and potential utilized.

The centrally sponsored Command Area Development (CAD) programme was initiated in 1974-75 with the objective of bridging the gap between the creation and utilisation of irrigation potential and for optimising production and productivity from irrigated lands on sustainable basis.

The programme mainly involves on-farm developmental works like construction of field channels and field drains, land lavelling and shaping and introduction of conjunctive use of canal and tubewell irrigation.

CAD programme is in operation in the country for more than two decades. During the 8th Five Year Plan Period, performance evaluation studies of irrigation projects were taken up. One of the major constraints of such evaluation studies is the absence of baseline inventory of irrigation projects on spatial scale with the Command Area Development Authority (CADA). The lacunae was felt most when the Ministry of Water Resources (MoWR), Govt. of India launched the National Water Management Project (NWMP) with the World Bank's assistance in early nineties in some of the irrigation projects. Since then, MoWR is pursuing evaluation studies with several alternatives including application of satellite remote sensing.

An opportunity came in early 1998, when MoWR and National Remote Sensing Agency (NRSA) signed agreement to take up satellite remote sensing based evaluation study of 14 major irrigation projects as a bench mark exercise to be completed in 2 years time frame.

These irrigation projects are : Sriramsagar and Nagarjunasagar (Andhra Pradesh), Jamuna and Sukla (Assam), Krishna, Jeyakwadi, Girna, Upper Tapi, Bhima and Purna (Maharashtra), Chambal (Rajasthan), DVC, Kangsabati and Mayurakshi (West Bengal) covering 3.3 million hectare culturable command area (CCA). This paper deals with Nagarjunasagar left bank canal irrigation command as one of the above-mentioned bench mark studies to realize application of spatial information technologies.

Nagarjunasagar Left Bank Irrigation Command in Andhra Pradesh, India Performance Evaluation Matrix using Spatial Information Technologies.

Fig.1 Cropping Pattern

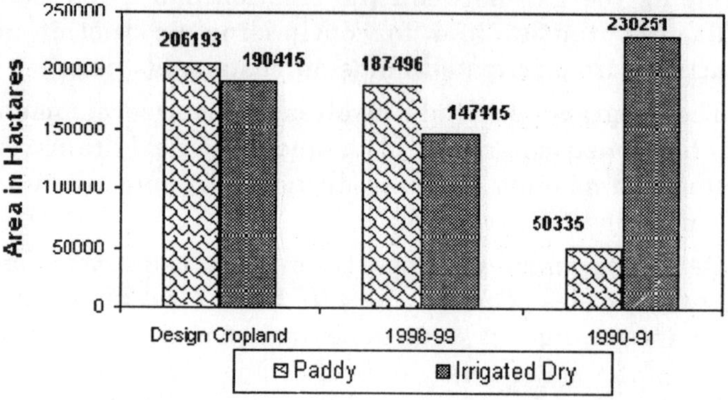

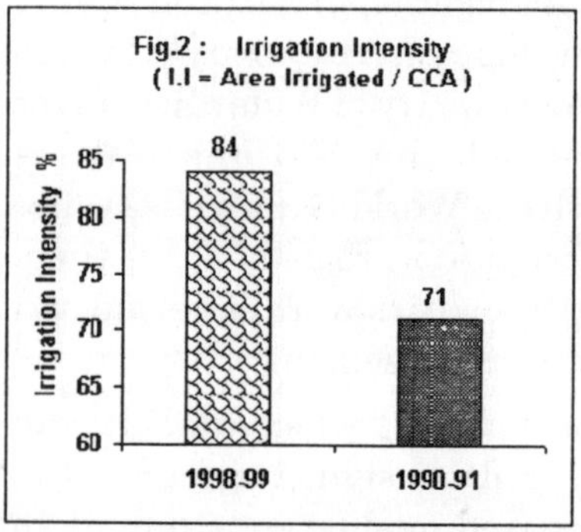

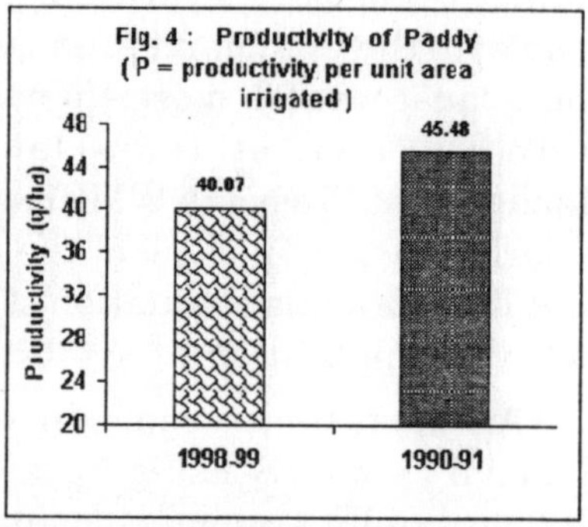

Fig.5 : Water Utilization Index
(WUI = Area Irrigated per unit of water supplied)

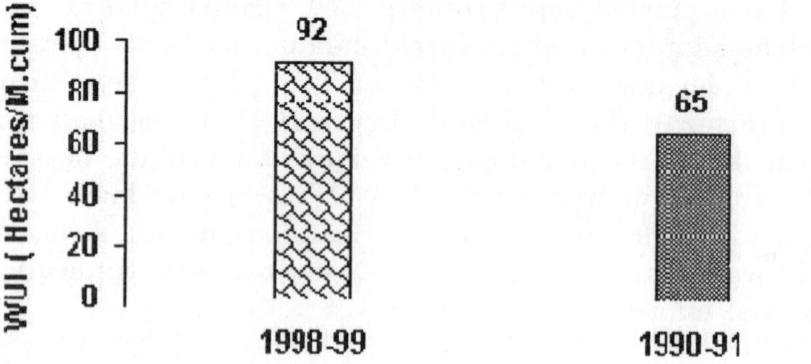

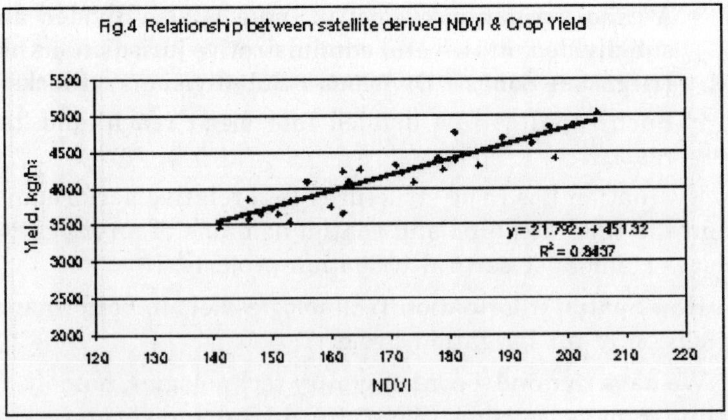

PERFORMANCE EVALUATION MATRIX

Basic objective of a irrigation project is to improve the productivity of the land for agricultural produce with scientific application of water and other management inputs. Performance can therefore be judged, if we have answers to the following questions:

- Is the irrigation potential utilized vis-à-vis the potential created?
- Is the return on agricultural production and productivity achieved towards design crop yield (maximisation)?
- Is irrigation water being efficiently applied (managed)?
- Is there an improvement or deterioration in the irrigated lands development (due to salinity / alkalinity, waterlogging)?

To address these questions, 'time domain' and 'spatial domain' play important role.

Time domain: Time span of performance evaluation can be (i) since commissioning of the irrigation project, and or (ii) in subsequent time periods.

Spatial domain: A canal irrigation project in India constitutes a heirarchial system of main canal, branch canals, distributaries, minors and field channels. Each has its irrigation command.

- A canal irrigation project in India is also divided and sub-divided into several administrative jurisdictions like Irrigation Zones / Divisions / Sub-divisions / Blocks.
- Each canal can be divided into 'head reach' and 'tail reach'.

Evaluation has to be structured into relative performance within the time domains and spatial domains of an irrigation project or amongst several irrigation projects.

How Spatial Information Technologies help in Performance Evaluation of an Irrigation Project?

We have in mind 4 contemporary technologies, namely, (i) satellite remote sensing (SRS), (ii) digital image processing (DIP), (iii) geographic information system (GIS), (iv) global positioning system (GPS) to apply to address this issue.

- *Satellite Remote Sensing*

Crop classification using multi-spectral data of IRS LISS is now well-established operational tool in India. Based on the crop calendar, optimal satellite datasets covering the entire crop season (say, for rabi crop season, one data set each month during November to March) are selected. Before classification of an image, few pre-processing steps, like geometric rectification of satellite data using ground control points (GCPs) and normalisataion of multi-sensor image data are followed.

- *Digital Image Processing*

Geometrically rectified multi-date satellite data is sequentially analysed with maximum likelihood classifier algorithm (in the present study, using Erdas Imagine ver. 8.3.1 in Silicon Graphics workstation platform) supported with ground-truth (training sets) collected during field visit. Various crops are identified and extracted sequentially at different stages of the classification process from the multi-date satellite data during the crop growth stage period. Classification accuracy with respect to training sets and total thematic map is ensured by close examination of error matrix (confusion matrix). Multi-date satellite data are also transformed to NDVI (Normalised Different Vegetation Index) values using the mathematical function of ratio of difference between reflected radiation in infrared and red band values by sum total of these two band values. NDVI represents the integral effect of various factors that influence crop production and is well accepted in India and elsewhere for crop yield forecast and agricultural drought monitoring.

- *GIS Application*

Usually, maps available with Command Area Development Authority (CADA) are of various formats and themes consisting of canal network, irrigation command boundary / administrative jurisdiction, place names, besides rivers / streams, reservoir / tanks, settlements, transportation networks, etc. A digital database is created of this information, which is co-registered with multi-date satellite data. When the crop classification maps and crop condition (NDVI) maps are available from satellite derived data, extraction of information within various spatial domains of the irrigation command can be done by overlaying digital database. GIS modules in Erdas software are used for these purposes.

- *GPS Application*

Satellited based GPS provides accurate geo-referenced (in terms of latitude & longitude) positional location on the ground. Stand-alone mode GPS instrument is used to determine the geo-reference position of CCE (Crop Cutting Experiment) plots in the crop fields, maintained by CADA for crop yield estimation. These CCE plots (usually 5 metre x 5 metre in size) are then identified onto the satellite image to extract NDVI values of the CCE plots. A regression model between NDVI and crop yield for the sample CCE plots is developed which is then extrapolated to the spatial domain(s) of the irrigation command for estimation of crop production.

NAGARJUNASAGAR IRRIGATION COMMAND

Nagarjuna Sagar Left Canal (NSLC) command is designed for irrigation of 0.42 million hectare CCA in 3 districts in Andhra Pradesh State. The spatial domain of NSLC consists of (i) 3 irrigation zones which is further sub-divided into 32 irrigation blocks (iii) located in 3 districts and (iv) is supported by 114 canal distributaries. Two time domains and one crop season, namely, rabi crop season (November-March) of 1990-1991 and 1998-1999 are selected to identify progressive performance in NSLC command. The methodology to use SRS, DIP, GIS and GPS to transform primary data (satellite image / CADA maps & office data) to secondary information (performance matrix) is followed, as mentioned earlier. All spatial (maps) and attribute (crop types, crop area, cropping intensity, and crop yield) information are generated with reference to 2 time domains and 4 spatial domains. In the following statements, the portion 'underline' indicates the performance Lmatrix in the NSLC irrigation command as a whole. Performance matrix with respect to disaggregated spatial units are cited in Ref.3.

- Crop classification from IRS 1A LISS II and IRS – 1C/ 1D LISS III data consists of paddy, cotton, pulses, tobacco, surgarcane, chillies and orchard crops. These crops are grouped into 'irrigated wet' and 'irrigated dry', as operationally followed in this command. Inconsistency in cropping pattern during the two time domains is observed (Figure 1). Irrigation service being extensive in 1998-1999 compared to 1990-1991, irrigation potential utilized is higher by 55,000 hectare and cropping intensity has increased from 71% to 84% (Figure 2).

* Paper presented in : 5th Annual International Conference, "Map India 2002", organised by CSDMS at New Delhi, 6-8 February 2002.

- A crop yield model for the major crop (paddy) has been developed by correlating satellite data based time composited NDVI values at CCE plot locations with the estimated yield in the CCE plots (Figure 3). The model provides crop yield in tonnes / hectare, which is an important performance indicator across space and time domains. The model is validated and is found to predict well at the larger disaggregated units. In the process of crop yield model development, satellite based GPS technique is used to determine the geo-referenced locational information of CCE plots in the satellite image. Although, irrigation service was extensive in 1998-99, paddy crop productivity (Figure 4) was found to be lower (4.00 tonne / hectare) compared to 1990-91 (4.55 tonne / hectare). However, these yield estimates are much above the national average from the irrigated croplands.

- Water release into the canals is utilized to calculate Water Utilisation Index (WUI). During 1990-91, area irrigated was much less for per unit of canal irrigation water supplied (65 hectare irrigated / million cu.m water supplied), compared to 1998-99, when better utilization of canal irrigation water was achieved (92 hectare irrigated / million cu.m water supplied).
- All the performance parameters, therefore, indicate general improvement in system performance of NSLC project.

2

Land Degradation due to Hydro-salinity in Semi-arid Regions

Fourth drainage project located in the center of Punjab province of Pakistan was selected as a research field. A newly introduced tile drainage system and remodeled surface drains have gradually reduced the waterlogging problem, however, salinity still remains in patches in the area. The present study describes an attempt to monitor land and water management scenarios that caused deterioration of vast productive land, using satellite data.

IRS-IB LISS-II digital data acquired were digitally analyzed along with other field data sets and topographical maps. False Color Composite and unsupervised classified image shows that most of the waterlogging and salinity problems are located along the drainage canals. The buffering analysis along the drainage and irrigation canals was performed to see the hydro-salinity impact on productivity in terms of its vegetation vigor. The Normalized Difference Vegetation Index (NDVI) tends to be lower along drainage and higher along the irrigation canals. It is probably because of the inequity water distribution and locational disadvantage of the down-stream farmers as well. A contrast example of an enterprise model-farm near the project area was observed which does not show any locational unbalance/ inequity but exhibits comparable high NDVI values almost all over the farm area, except fallow lands. The extent of waterlogged soils was also estimated through GIS analysis from more than 150 peizometers locations lying within the study area. It was found that 24.5%, 22.0% and 63.4% of the mapped area was under risk of waterlogging (200cm) during 1990, 1993, and 1996 respectively. Evidence is presented that the existing drainage system along with use of groundwater for irrigation reduces the extent of waterlogged soils. However, the reuse of poor quality water to supplement irrigation supplies by the downstream farmers and the failure of a few drainage sumps are likely to disturb the water balance resulting into increased risk of waterlogging and salinity in the area.

INTRODUCTION

Environmental land degradation has recently become a global, urgent issue and is now being considered with high priority, specially in the developing counties, to meet food and fiber demands of accelerated population pressure with the limited available resources. Secure balance in the global supply and demand for food has forced humans to develop agriculture in semi-arid to arid lands, which are generally less suitable for agriculture and sensitive to environmental changes. Irrigation systems have always been considered as an effective way of increasing agricultural production since ancient times to bridge the gape of water shortage. Owing to irrigation systems, the farmers have obtained quite stable productivity but they started facing an acute problem of waterlogging and salinity due to seepage from the huge system in most of the developing countries like Pakistan. The farmers, many local and international organizations are struggling now against the so-called "white death" of their lands. Land degradation due to waterlogging and subsequent salinization has been so enlarged that it is now being regarded as a global environmental problem, desertification. It is, therefore, important to monitor land and water management scenarios causing severe land degradation and low productivity. Remote sensing is one of the key tools in monitoring local, regional and global environmental issues. More recently, much attention has been paid to spatial analysis due to merging of geographic information system (GIS) and satellite images for environmental research and applications. The conventional means are however, not only difficult and time consuming but also laborious due to vagaries of the weather. It is prudent to use such emerging technique with an emphasis to its application in semi-arid areas. The present paper describes an attempt, where in IRS-1B, LISS-II (Indian Remote Sensing Satellite) digital data have been used along with other maps and field data, to monitor land degradation due to hydro-salinity and low agricultural productivity (vegetation condition).

Research Field

A reclamation project called as "FDP" (Fourth Drainage Project) was launched to mitigate the twine menace of waterlogging and salinity, in 1983. The FDP area is located in the center of Punjab province of Pakistan and consists of Schedule-I and Schedule-II units. The Schedule-I area covering about 70,000ha was selected as a research filed, lies between 31° 02' to 31° 45' N and 72° 50' to 73° 22' E (Figure 1). It was estimated that 77% and 43% of the total area was under high watertable conditions and salinity respectively, before the installations of a new tile drainage system (Table 1). Mean annual rainfall of the locality is <350mm out of which nearly 75% occurs during monsoon season (July- September).

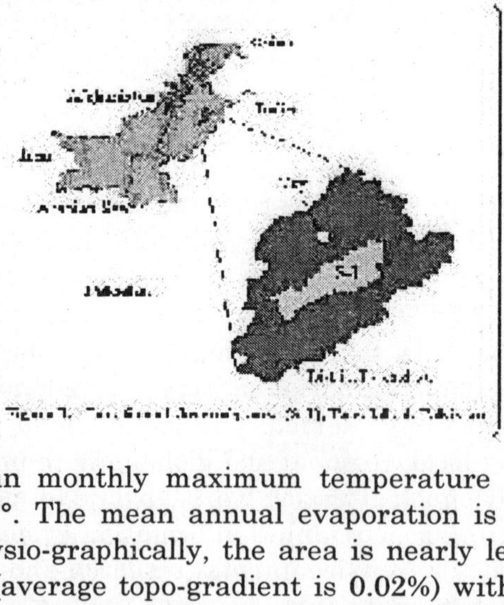

The mean monthly maximum temperature is between 19.4 - 41.2 C°. The mean annual evaporation is as high as 2100mm. Physio-graphically, the area is nearly level to very gentle slope (average topo-gradient is 0.02%) with mean sea level of 190m. Geologically, the soils are alluvial deposits classified as silt loam, loam, and silt clay loam and loamy sands. There are two cropping seasons i.e. Rabi (winter) and

Kharif (summer). Main Rabi crops are wheat, sugarcane, pulses, and fodder while corn, paddy, cotton, sugarcane, and fodder occupy lands in kharif season.

Table 1. Waterlogging and salinity situation in the Fourth Drainage Project in January 1983.

Status	Salinization			Watertable Position		
Area (%)	Salt Free	Saline Sodic	Saline-Non Sodic	0.00-1.5m	1.50-3.0m	>3.0m
	57 %	35 %	8 %	77%	21 %	2 %

Data Used

Two types of data have been used in this study, i.e. satellite and field data. Satellite data of IRS-IB, LISS-II (linear image self-scanning spectrometer) in all four bands (B1: 0.04-0.52um, B2: 0.52-0.59um, B3: 0.62-0.68um and B4: 0.77-0.86um) are used for three time periods of year1993.

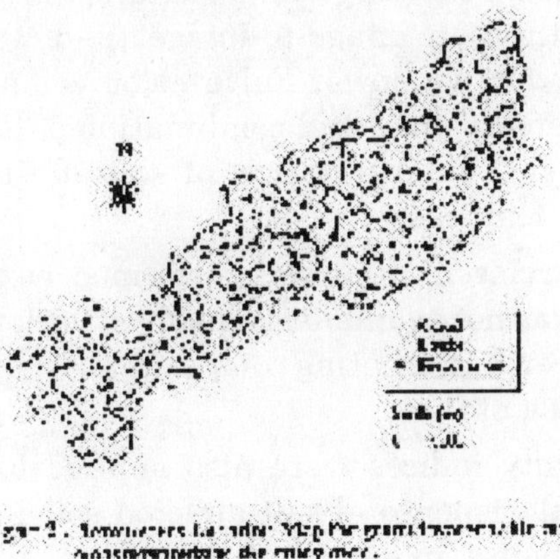

[illegible]

Three different months have been chosen for the satellite data because of considerable variation in salinity as well as in vegetation cover. Topographic maps of survey of Pakistan used in the study were 44 E/3, 44 E/7, 44 E/8 and 44 /16 of scale 1:50,000. Layout maps of irrigation canal and surface drains were also used. The watertable data for more than 150 Piezometer locations (Figure. 2) lies within the S-1 area were used to know the status of soils under waterlogging risk. ILWIS 2.2 for Windows system has been used for all remote sensing and GIS analysis. Information about water and soil quality, soil types, agricultural practices, crop yield, irrigation facilities and necessary ground truth data of the study area will be obtained and used in the remaining part of the study.

Study Approach

The IRS LISS-II digital data were registered to Survey of Pakistan topographical sheets using an image-to-map registration algorithm (UTM map projection with 36.25m ground resolution). About 30 common, well identified and uniformly distributed GCP points were selected from both, the displayed image and topographical sheets.

The image was then registered using affine transformation method and the nearest interpolation was used for arrangements of the coordinates of each picture element. The average error was within one picture element in terms of precision after the geometric correction. Following this procedure, the others scenes were registered through image-to-image tie-down algorithm using ILWIS 2.2 for windows. False color composite (FCC) image was generated using band combination of R:G:B = 4:3:2 for better and visual interpretation of salt-affected lands in contrast to other features.

Unsupervised classification was performed, as the complete site information was not available. It involved various processes like image enhancement, creating composite files, using cluster/iso-cluster modules etc.

Various salinity indices were also applied based on the concept of spectral response of salt-affected soils. It is to note that spectral response in terms of digital no. (DN) of salt-affected soils is relatively higher than other categories in band-1 (B1) and band-3 (B3). The following two salinity indices namely S.I (Salinity Index) and NDSI (Normalized Differential Salinity Index) proposed by Tripathi et. al. (1997) were applied which relatively give good results in re-classification of salt-affected lands. NDSI is just the reverse of NDVI index for vegetation.

S.I = (Band 1 X Band 3) ½ (1)

NSDI = (Band 3 - Band 4) / (Band 3 + Band 4) (2)

The strongly visible salt-affected lands were then extracted from the final classified image to make a thematic map. The sum of all the extracted salt-affected soils was prepared, considering the rest of area as slightly saline to normal lands (Figure 4). Depth to watertable data for the year of 1990, 1993, and 1996 for more than 150 piezometer, reported by JIID (1997) along with their GPS locations, were used to see the spatial watertable positions in the area.

Using moving surface interpolation technique, initially the average watertable positions maps were prepared and analyzed for each year and then a map of "mean watertable positions" based on the three year results, was prepared. It is further grouped as waterlogged and non-waterlogged conditions, assuming an assumption of critical watertable depth of _ 200m for the area (Figure 5).

Reflectance variations of vegetation on the image are attributed to the different species of vegetation and their densities, which together provide evidence of shallow ground watertable conditions and saline agricultural areas. Favorable growth conditions prevail in regions where the watertable is situated below the area of influence of evapotranspiration, that is, within 10-m depth (Srivastava 1997). An indication of whether scanty vegetation in an area is due to high watertable depth or salinity can be investigated using Normalized Differential Vegetation Index (NDVI) which easily grasp the state of vegetation. The NDVI for the area can be expressed as:

NDVI = (NIR Band _ Red Band) / (NIR Band + Red Band) (3)

The maps of irrigation canals and surface drainage network were digitized in the GIS environment. Some discrepancies of mismatching were found when overlaid on the geo-referenced image. These layers then were corrected with the help of topographic maps and image. It was presumed that the topographic sheets could be the most reliable source of information. Buffering analysis along canal and drain channels was performed for 1000m corridor to see the hydro-salinity impact on productivity in terms of vegetation vigor.

Result and Discussion

Waterlogging and Salinization The IRS composite image was systematically visually interpreted with the help of old salinity maps/ information developed in early 80's. It is revealed that salt-affected and waterlogged areas are clearly delineated compared with other land use features in FCC pellet. The most of the salt-affected lands were observed more pronounced near the drainage network as shown in white to light bluish tones. JIID (1997) and Casas (1995) reported similar results.

This could be best explained, as the watertable along the drainage canals has found shallower relatively due to low location. The irrigation canals are constructed on relatively high locations comparing to the drainage system to utilize the natural gradient effectively. The occurrence pattern of salt-affected soils through unsupervised classification method was grouped into two main classes i.e. slightly to normal saline and visible salt-affected soils. The existence percentage of salt-affected soils was found to be 7.5 % for the mapped area of S-1 unit. Map of 'mean' depth of watertable shows the extent of waterlogging (area _ 200cm) in the area. The yearly average watertable analysis indicates the lowering trends during 1990-93 from 24.5 % to 22. %. However, it started increasing tremendously by 39 % in 1996. This situation is quite alarming when comparing the change of watertable for the 7 years period gape. Low with-drawl from the sumps, low natural drainage to the aquifer, deterioration of many vertical drainage system (tube-wells), choking of various sub-surface drainage lines by debris like weed growth and silting have appreciably contributed to the rising of watertable (WAPDA, 1997). All other sources of recharge and discharge of groundwater reservoir are, more or less, the same except the unpredictable rainfall pattern but it hardly cause any big change in the hydraulic balance of the area. Many farmers were reported for re-use of poor quality drainable surplus to supplement irrigation needs. Most of the farmers complained water shortage in the socio-economic survey

(WAPDA/SMO, WM&E, 1993-94) for the whole area, especially, big percentage for the tail-end users. Inequity water distribution were also reported among the shareholders of a watercourse at the head, middle and tail reach.

AGRO-PRODUCTIVITY IN RELATION TO HYDRO-SALINITY.

The results of the vegetation index (NDVI) were analyzed along the irrigation and drainage channels (the area under easy risk of hydro-salinity) indicated rather scarcity of vegetation as a whole. Relatively healthy growth of the green bio-mass were observed along the canals than that along the drains. Easily visible high NDVI were noted for a private agricultural model farm (Sehgal model farm) near the FDP area for all periodic NDVI images, except fellow lands. It is due the fact that the water management is demand oriented there (3-times more water allocation i.e. 4.7mm/d than the average of FDP area as 1.54mm/d), thus more flexible with sufficient water supply. The design irrigation allocation (1.8-2.1mm/d) for the area is half of average crop water requirement (JIID, 1997). These results clearly show the possibility of resolving the remaining salinity problem out of area. The NPP (Net Primary Productivity) for natural vegetation corresponding to the annual precipitation range of about 300mm was too small and quite sensitive to changes in the rainfall pattern (JIID, 1997). As NPP is closely related to the agricultural productivity in rainfed farming, it is certain that ample irrigation supplies are necessary to obtain sustainable agriculture in the area.

Composite image and unsupervised classified image shows that most of waterlogging and salinity problems are located along the drainage canals. The existence percentage of salt-affected soils was found to be 7.5 % for the mapped area of S-1 unit. Th situation of waterlogging is quite alarming when comparing the change of watertable for the 7 years period with the lowering trends from 24.5% to 22% (1990-93) and clear increase by 39% in 1996. Evidence is presented that the existing drainage system along with use of groundwater for irrigation reduces the extent of waterlogged soils. However, the reuse of poor quality water to supplement irrigation supplies by the downstream farmers and the failure of a few drainage sumps are likely to disturb the water balance resulting into increased risk of waterlogging and salinity in the area. The vegetation growth analyzed through NDVI tends to be lower along drainage and higher along the irrigation canals due to the inequity water distribution and locational disadvantage of the drains and vice versa. High NDVI values for almost all over the model farm near FDP project area with demand oriented water supply thus gives the solution to resolve the remaining salinity problem. Lastly, it is concluded that the water shortage is one of the fundamental problem in the area to reclaim the soils for sustainable productive lands.

3

IRRIGATION PERFORMANCE OF RICE-BASED BHADRA PROJECT IN INDIA

The study in Bhadra Project in India is an attempt to use a package of satellite remote sensing applications to assess agricultural system performance of rice-base irrigation system. Multi-year satellite data have been analyzed to provided disaggregated information on the irrigated area, cropping pattern and rice productivity. This spatial and temporal information has helped in evaluating the agricultural system performance through the years and across the helped in evaluating the agricultural system performance through the years and across the irrigation scheme. Problem pockets with limitations with regard to irrigation intensity, rice productivity and equity have been identified for follow-up action toward improved performance. This irrigation scheme, after rehabilitation under the National Water Management Project (NWMP), has shown significant improvement in the extent of irrigation as well as in the agricultural productivity.

The unit cost of SRS application in this irrigation scheme of 100,000 size works out to 10 develop appropriate vegetation index which is optimally sensitive to rice characteristics rather than to the background as well as improvement in the spectral modeling of rice yield.

Introduction

Performance assessment has been prioritized as most critical element to improve irrigation management (Abernethy and Pearce, 1987). Though remote sensing was identified as a tool to assess performance more than a decade ago (Abernethy and Pearce, 1987), the actual applications have been very few so far.

Many early applications focuses on mapping irrigated crop lands (Huston and Titus, 1975; Draeger, 1976; Wall, 1979; Thiruvengadachari, 1981) and have continued till recent years, though with increased inventory capabilities with regard to irrigated area from different sources (Thiruvengadachari, 1983), crop type and crop stress (Azzalli and Menenti, 1989) and to monitoring temporal changes in irrigated area (Nageswara Rao and Mohan Kumar, 1994).

This study in Bhadra Project in India is perhaps the first attempt to use a package of satellite remote sensing applications for desegregated inventory through a number of years of irrigated area, cropping pattern, and estimating major crop (paddy) productivity to develop agricultural system performance indicators. Diagnostic analysis of selected poorly performing pockets of command area complemented the performance assessment. The study involved application of known procedures in an operational context as well as extension and innovative improvement of these techniques to address field complexities.

Bhadra Project

The Bhadra Project is located on the Bhadra River, a tributary of the Krishna River, in the state of Karnataka, India. The project comprises a dam with a gross storage capacity of 2025 Mm3, a Left Bank Canal serving 8290 hcctares (ha) and a Right Bank canal serving 92,360 ha. The Bhadra dam is situated 50 km upstream of the point where the Bhadra River joins Tungabhadra, another tributary of Krishna, and intercepts a catchment of almost 2000 km2 (Figure 1).

The long term average annual precipitation is 827 mm. Although the Bhadra basin gets rainfall during both southwest and northeast monsoon period. Sixty percent of the command is under red soil while the balance is under block soils. The mean annual potential evapotranspiration has been estimated as 1678 mm.

Figure 1 Location of the Bhadra Project command area

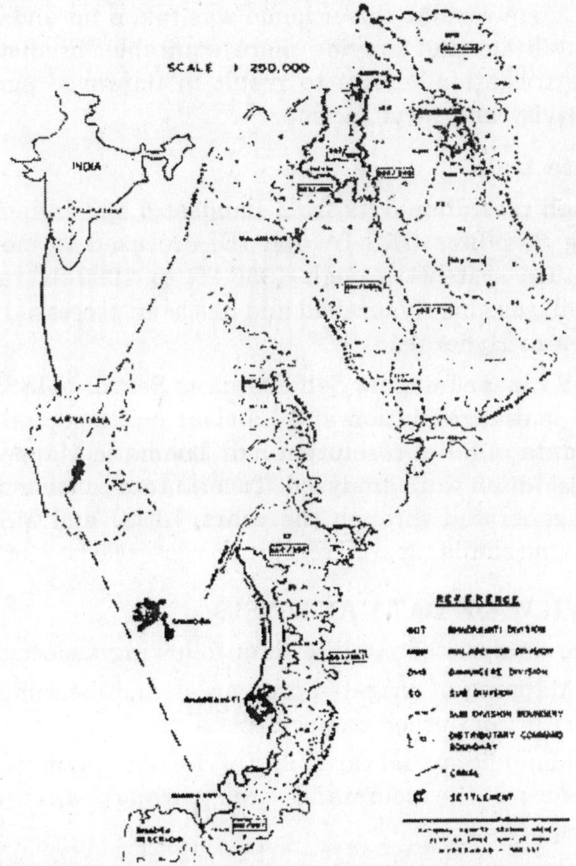

The project was designed to irrigate semi-dry crops that were to occupy more than 60 percent of its command area, with an overall annual cropping intensity of 200 percent. But as agricultural development progressed, it was found that ponded rice dominated to the extend of occupying 90 percent of the irrigated land on the Left Bank and about 56 percent on the Right Bank.

The heavy demand for rice led to inequitable use of the irrigation supply, and resulted in the rapid deterioration of the irrigation system as farmers intervened to modify the water management plan. This not only threatened the physical collapse of the system but also provoked dissatisfaction among till-end farmer. As a results, this scheme was taken up under NWMP to rehabilitate and provide more equitable, predictable and reliable irrigation service to result in improved agricultural productivity and farm income.

Salellite Data

High resolution data from Landsat 5 and Indian Remote Sensing Satellites (IRS IA and IB) ere used to monitor the agriculture system through 1986-87 to 1993-94 rabi (post-monsoon) seasons Hyderabad and has been processed at NRSA facilities in Hyderabad.

IRS Linear Imaging Self Scanning Sensor (LISSI) data of 72.5 m spatial resolution and Landsat multispectral Scanner (MSS) data of 80 m resolution and Thematic Mapper data of 30 m resolution were analyzed. To maintain consistency in the results generated through the years, LISSI and MSS sensor data having similar spatial resolution were selected for analysis.

OVERVIEW OF DATA ANALYSIS

The analysis of satellite data following aspects:

- Mapping of irrigated crop areas and discrimination of rice from other cropped areas.
- Mapping spatial variability of rice transplantation-period across the command area through an innovative approach.
- Rice yield estimation through spectral index-yield models.
- Improved design for selection representative sample areas where yield data are measured based on satellite derived data on rice condition.
- Evaluation of impact of reported waterlogging on rice productivity.

A Geographic Information System (GIS) was developed at two levels of command area and for selected distributors fir evaluating the system performance and to diagnose and analyze the poorly performing distributors.

The Bhadra irrigation system is characterized by major area under ponded rice followed by semidry crops, sugarcane and garden crops and rice transplantation is also staggered over a period of more than a month and semi-dry crops being sown considerably earlier to rice. In view of this heterogeneity in crop calendar, both in order to obtain complete estimate of area under any crop as well as to ensure better discriminability, satellite data of 2 dates one at the time of maximum ground cover and canopy growth and another earlier data when paddy is being transplanted with semi-dry crops already sufficiently grown, during the irrigation season were analyzed. After a review of alternate classification approaches, a multivariate classification approach with two dates data merged into a single multi-channel data set was attempted. To maintain acceptable accuracy even at distributary level, final classification consisted of only rice and non-rice crops. The post classification check through field visits in more than 300 randomly selected points validated the classification accuracy to be 90 to 95 percent. Distributary level crop area statistics were extracted by digitally overlaying the base maps of the command area on geometrically rectified crop classification map.

Spatial information on transplantation time for rice across the command area has been mapped. The seasonal NDVI (defined as the ratio of difference in red and near infrared reflectance to their sum) profile of every pixel of rice crop was analyzed to identify the peak-greenness stage. The knowledge that this corresponds on spatial staggering of rice transplantation. This is useful for evaluating the compatibility between canal delivery schedule and rice crop calendar at the distributary level.

In the Bhadra Project study, normalized Difference Vegetation Index (NDVI) has been used to correlate with rice yield. A simple rice yield model was developed based on relationship between peak NDVI at heading stage of crop to the yield at 72 plots obtained through crop cutting experiments during 1992-93 rabi season level.

In the Bhadra Project study, normalized Difference Vegetation Index (NDVI) has been used to correlate with rice yield model was developed based on relationship between peak NDVI at heading stage of crop to the yield at 72 plots obtained through crop cutting experiments during 1992-93 rabi season.

Since rice transplantation is staggered across the command area, satellite data of any one data does not represent the same growth stage at all locations. An innovative approach of time composition was attempted, in which the maximum NDVI value for each rice pixel was picked from the satellite overpass enveloping paddy transplantation period. NDVI is the highest at the heading stage of rice crop. The rice yield model is defined as yield (kg/ha) = 42.23 TCVI -3439, where TCVI-time compo sited NDVI derived from IRS LISSI data. The TCVI for every pixel is the maximum NDVI value reflected from coregistered multidate satellite data enveloping the heading stage. The fractional maximum NDVI is multiplied by 400 to transform into the dynamic range of 0 to 255. The yield model developed from satellite and ground data of 1992-93 rabi season has been validated during 1993-94 rabi season, and the maximum deviation from observed crop cutting experiment yield values was less than 10 percent indication the stability of the yield model. The rice yield model is currently location and time specific, since difference in rice variety, atmospheric effects and fertilizer-vegetative growth relationship may require modification of model coefficients.

SYSTEM PERFORMANCE EVALUATION

Primary data on agricultural productivity (cropping pattern and rice yield) have been generated at distributary and reach levels from satellite data for rabi seasons from 1986-87 to 193-94 (Tables 1,2 and 3) and the improvement on system performance after NWMP implemnetation has been evaluated (Table 4). The tables indicate that irrigation intensity has increased in every sector of the command area, and distributarties with less than 50 percent irrigation intensity are almost negligible. Also, rice production has grown not only in terms of cultivated area, but also in yield level and production. Although rice was to be precluded and semi-dry crops encouraged during the rabi season under the NWMP, the case study found that irrigation intensity during the rabi season increased from 76 percent to 91 percent, rice area from 56 percent to 69 percent, and rice yield from 3.8 tonnes/ha to 4.9 tones/ha during the above period.

Table 1. Irrigated crop area (ha) through the years.

Division	Planned Command(ha)	Irrigated Area (ha.) 1986-87	1987-88	1988-89	1989-90	1992-93	1993-94
Davangere	45,280	33,838 (75)	15,848 (35)	31,760 (70)	31,760 (69)	31,038 (87)	39,436 (84)
Malebennur	35,030	28,242 (81)	23,260 (36)	23,260 (66)	31,738 (91)	31,738 (91)	30,610 (87)
Bhadravathi	17,050	11,449* (67)	5,856* (34)	13,945* (82)	13,311* (78)	17,250* (101)	15,695 (92)
Total	97,360	73,529(76)	34,381(35)	68,965(71)	67,366(69)	88,424(91)	84,412(87)

*As per ground data; all others are satellite based.
Note : Figure within brackets represent irrigation intensity in percdcent.

Table 2. comparison of percent rice area during the rabi season

Division / Subdivision	Percentage of rice area 1986-87	1988-89	1989-90	1992-93	1993-94
Subdivions 1	45	39	23	51	60
Subdivions 2	54	49	40	67	69
Subdivions 3	59	57	44	78	78
Subdivions 4	42	37	35	63	74
Davangere Division	51	46	35	66	71
Subdivions 1	72	83	90	88	83
Subdivions 2	56	33	46	66	54
Subdivions 3	59	61	60	73	75
Subdivions 4	61	42	79	77	72
Malebennur Division	65	64	76	81	76
Subdivions 1	60*	53*	21*	61	64
Subdivions 2	41*	33*	33*	61	53
Subdivions 3	36*	33*	34*	49	30
Bhadravathi Division	48*	51	46*	58	49
Total command	56	53	51	69	69

* As per ground data.

Table 3. Rice (rough rice) yield through the year

Division/Subdivision	Yield (tones/ha)			
	1986-87	1989-90	1992-93	1993-94
Subdivions 1	2.4	4.5	3.8	4.2
Subdivions 2	3.7	4.7	4.1	4.2
Subdivions 3	3.4	4.2	3.8	4.6
Bhadravathi division	3.2	4.6	4.0	4.3
Subdivions 1	4.3	6.1	5.8	56
Subdivions 2	3.3	5.0	4.0	4.5
Subdivions 3	4.1	6.0	4.9	52
Subdivions 4	3.7	5.7	4.9	52
Malebennur Division	4.1	5.9	5.3	5.3
Subdivions 1	3.8	4.9	4.0	4.8
Subdivions 2	4.0	5.2	4.5	4.6
Subdivions 3	3.7	4.5	4.4	4.4
Subdivions 4	3.4	5.2	4.5	4.9
Davangere division	3.8	5.1	4.4	4.6
Total command	3.8	5.4	4.7	4.9

Notes: 1986-87 Rabi pre-NWMP and no cutoff.

1989-90 Rabi with 25 percent command area cutoff from irrigation

1992-93 Rabi without cutoff.

1993-94 Rabi without cutoff.

The projects performance before and after NWMP implementation in 1988 is shown in table 4. Other performance indicators such as depth of water application and rice production per unit of water can be derived from satellite data in conjunction with ground. The water delivery is measured at the off take point of the distributary. The system losses in the distributary have not been accounted for. The equity of the water supply is measured through Christiansen's Uniformity Coefficient (UCC) which is estimated as the ratio of the weighted sum of deviations of the depth of water supply from the mean and the mean depth of water supply (Table 5).

Table 4. System performance under NWMP, Bhadra project

Parameters	Rabi season 1986-87	1989-90	1992-93	1993-94
Irrigated crop area (ha)	73,529	67,366	88,424	84,41
Percentage of rice area	56	51	69	69
Average rice (rough rice) yield (tones/ha)	3.8	5.4	4.7	4.9
	1.06	1.04	0.80	0.86
Depth of water application (m/ha)	0.94	0.96	1.25	1.16
Area irrigated per unit volume of water	3.8	5.4	4.7	4.9
(ha/ha.m)	0.3	0.4	0.5	0.5
Rice (rough rice) output per unit of land (tones/ha)				
Rice (rough rice output per unit volume of water (tones/m^3)				

Table 5. Equity in water supply during the 1993-94 rabi season

Branch Canal	Christiansen's Uniformity Coefficient among: Sub division	Distributaries: Sub-distributary 1	Sub-distributary 2	Sub-distributary 3	Sub-distributary 4
Bhadravathi	0.75	0.91	0.80	0.75	-
Malebennur	0.93	0.85	-	0.69	-
Davangere	0.919	0.66	0.84	0.93	0.89

The performance of the Bhadra Project through the years has shown significant improvement in regard to area irrigated, agricultural productivity and equity of distribution. Only a few pockets of inequity remain to be improved by the management.

COST EFFECTIVENESS OF SRS TECHNIQUES

Satellite remote sensing application to disaggreated irrigation inventory in Bhadra Project of 100, 000 ha size costs about 10 U.S. cents per ha for each irrigation season at 1994-95 price level. This includes the cost of satellite data, analysis and statistics generation. The unit cost decreases to about 3 U.S. cents per ha for irrigation schemes larger than 250,000 ha due to economy o scale. Similar unit cost for monitoring waterlogging and soil salinity status work out to 5 U.S. cents per ha and 2 U.S. cents per ha respectively, when attempted along with crop inventory. The SRS application cost works out to about 1 to 2 percent of annual operation and maintenance costs for irrigation schemes in India.

The Bhadra study has demonstrated the potential and cost-effectiveness of SRS techniques for inventorying and monitoring agricultural productivity in a large irrigation system in India. Effective integration of GIS with SRS techniques further enhances performance evaluation and diagnostic analysis capabilities. The study also points out the need to focus on development of appropriate vegetation index for rice and improved spectral yield models.

SYSTEM PERFORMANCE EVALUATION AND DIAGNOSTIC ANALYSIS OF CANAL IRRIGATION PROJECTS

Satellite remote sensing techniques can efectively complement and supplement ground data collection towards objective and reliable evaluation of irigation system performance and diagnosti analysis. Multidate satellite data during the rabi seasons prior to and after implementation of National Water Management Project NWMP have been analysed to provide spatial information on irrigated area and paddy productivity right up to distributary command level add through the years since 1986-87 rabi season. The Bhadra project performance is seen to have significatntly improved after NWMP in terms of increased irrigated area, improved paddy yield, decreased depth of water applied and improved water use efficiency with respect to paddy production. Diagnostic analysis has indicated problem distributaries of gap in irrigation utilisation, less nad stagnant paddy yield and poor water use efficiency. Equity in water application has also been diagnosed. Distributaries with large gap in paddy yield between head reach ad Satellite remote sensing applications are thus seen to be effective tools for irrigation water management.

The Bhadra rrigation Project in Karnataka State is one of the subprojects of National Water Management Project NWMP of India. The NWMP objective is to increase the agricultural productivity and farms income inexisting irrigation schemes by providing predictable, reliable and equitable irrigation service. The productivity would be measured in regard to agricultural yield and area irigated. The project incorporates monitoring and evaluation at various levels to provide information on system performance evaluation and diagnostice analysis, necessary for corrective measures. Satellite remote sensing techniques have been employed to effectively complement and supplement the ground mechanism in this project.

System performance indicators are depth of water applied irrigation intensity, major crop yield and water use efficiency. Equity of irrigation service has also been evaluated. Diagnostic analysis is carried out using temporl and spatial information at distributor level, Problem distributaries have been identified in regard to irrigation gap, low crop yield, low/no yield improvement through years and low water use efficiency. While many earlier studies have demonstrated the utility of satellite data in providing crop area statistics and crop yield assessment (Groten 1993, Rasmussen 1992) Tennakoon et all 1992, Thiruvengadachari 1981, Teng 1990) Over large administrative units, the investigation in Bhadra Project perhaps is the first attempt to provide informaton on spatal and tempra varability of performance indcatiors at distributary comand leve (Thiruvengadachari et a 1994.)

BASIC DATA AND METHODOLOGY

Multidate Indian Remote Sensing Satellite (IRS) LISS I data during the rabi seasons since 1986-87 have been analysed

(LANDSAT MSS/TM data to a limited extent has been used to fill tbe gaps in temporation coverage/non availability of IRS data) Kharif season could not be analysea in view of continuous cloud cover preciuaing satelite vrew of the ground.

Liss I sensor data was used since even at the spatial resolution of 72.5 meters this data could provide distributary level information on major crop groups and is also cost/time effective.

The satellite data has been digitally classified into paddy and non paddy crop areas. Though the latter has been further clasificed into individual crop types such as groundnut, sugarcane and garden crops, these were later lumped into one category, in view of reduced classification accuracy resulting from non contiguous smaller patchiness of such crops. Different classification schemes, ranging from single date to multidate data analysis and use of vegetation index along with raw bands, were attempted to provide the best delineation of irrigated crop area.

A base map of the command area, defining the distributary command boundaries, is prepared from survey of India topographic map in 1:50000 scale and revenue survey map in 1:7920 scale, taking the aid of topography, drainage network and other features. The base map is digitised and overlaid on the classified crop map, to enable crop statistices extraction under each distributary.

The crop condition is indicated by the normalised difference vegetation index (NDVI) which is an arithmetic combination of reflected radiation in the visible and near-infrared wavelengths of IRS/LANDSAT sensor.

The satellite data analysis is supported by field visits and ground data. The paddy yield model has been developed based on crop cutting experiments conducted during 1992-93 rabi season in the command area and NDVI of paddy.

Though irrigated crop area and yield as well as depth of water and water use efficiaency have been estimated at distrubutary level, only summary statistics are provided in subsequent sections.

SYSTEM PERFORMANCE EVALUATION

Irrigated Crop Area

Irrigation is being provided to a significantly larger area after NWMP implementation (Table 1) . The irrigation intensity in the current rabi season is 87 percent compared to 76 percent in the pre NWMP 1986-87 rabi season. Irrigation utilisation is highest in head end Bhadravathy division followed by Malebennur and Davangere deivisions. The tail end Davangere division has registered an increase of 9 percent in irrigation intensity. Aalysis of irrigation intensity at canal subdivision and at distributary level idicates that the irrigation intensity is high in No. 3 subdivision of Bhadravathi division.

The intensity is decreasing towards the tail end of Malebennur division, following similar trends before and after NWMP. The trend of decreasing irrigation intensity is seen in Dvangere division also. In general, while irrigation intensity has increased in recent rabi seasons, the spatil, while irigation intensity has increased in recent rabi seasons, the spatial variability is still following the pre NWMP trend. This non uniformity in irrigation intensity needs to be narrowed and equity established. Redistribution of water from distributaries with larger than 100 percent irrigation intensity can lead to greater equity.

The change in cropping pattern before and after NWMP is given in Table 2. The percent paddy area in the command area has increased from 56 percent in 1986-87 rabi and stabilised at 69 prcent in the recent rabi seasons. The percentage area under paddy crop is 49 percent in Bhadravathy division and around 76 percent in Malebennur division. Percent paddy area in Davangere division however is showing a significant increase in the recent two rabi seasons, perhaps indicative of improved water availability in this tail end division. Significant acroce is also feen in the tail end most No. 1 sub-division within vangere division. The spatial pattern of percent paddy area in cent rabi seasons is remarkably similar to the variability in 1986-87 rabi season, indicating alround farmer preference for paddy crop with improved irrigation service.

Table 1. Irrigted Crop area (ha) Through the years

Division	Planned Command	Rabi 1986-87	Rabi 1992-93	Rabi 1993-94
Davangere	45280	33838(75)	39436 (87)	38107 (84)
Malebennur	35030	28242 (81)	31738 (91)	30610 (87)
Bhadravathi	17050	11449 (67)	17250 (101)	15695 (92)

*As per ground data

Note Figure within bracket represents irrigation intensity

TABLE 2 COMPARISON OF PERCENT PADDY AREA DURING RABI SEASONS

Division	Rabi 1986-87	Rabi 1992-93	Rabi 1993-94
Davangere	51	66	71
Malebennur	65	81	76
Bhadravathi	48	58	49
Total Command	56	69	69

*As per ground data

Paddy Productivity

While conventional crop cuttin experiments have been conducted for the major crop of Paddy since 1991-92 the sampling scheme is often inadequate and not representative of the yield variability in the command area. Further the conventional scheme provides yield estimates only at the project level, and thus spatial variability between distrubutaries is not available nor such estimates are available for pre-NWMP years.

A bold approach of relating satellite data derived NDVI to paddy yield has been attempted. While earlier studies have demonstrated this relationship (William 1990 , Groten 1993) this is perhaps the first attempt at extending the NDVI yield model through years (after satellite data normalisation) and across distrubutaries in the command area.

Yield (kg/ha)= 42.23 TCVI-3439.05

Where TCVI is the time composited NDVI for any paddy pixel. TCVI is the maximum NDVI value for each pixel from the coregistered NDVI images of the three compeass dates.

The yield model has been developed using crop cutting experiments in 1992-93 rabi season at 70 plots, and has been validated through farmer surveys and CCE results of 1993-94 rabi season. Spatial yeeld estimates over canal subdivisions and distributary commands have been made usin the model. Temporal variability in paddy yield is obtained through application of radiometrically normalised satellite data in different years.

The satellite estimate of paddy yield in 1993-94 rabi season is given in Table 3. The average yield is estimated to be 48.71 Qtls/ha,with highest yield coming from Malebenur division (53.4Qtls/ha) followed by Davangere (46.4 Qtls/ha) and Bhadravathi division (42.9 Qtls/ha). The yield in Bhadravathi division is 12 percent lower than the weighted average yeild over the command area.

In Malbennur division, the paddy yield in No. 2 canal subdivision is 16 percent less than the divisional average yield. In Bhadravathi and Davangere divisions the subdivision yields are more uniform (within 10 percent of divisional average yield). There is a general decrease in paddy yield in te tail end under Bhadravathi main canal. The paddy yields are more uniform under Davangere branch canal.

The paddy yields under LBC and RBC of D.B. Kere pickup are 45 Qtls/ ha compared to the Malebennur division average of 53 Qtls/ha.

Table 3 Paddy Yield (Qtls/Ha) Through The Years

Division	Rabi 1986-87	Rabi 1992-93	Rabi 1993-94
Bhadravathi	32.18	39.67	42.88
Malebennur	41.10	52.89	53.38
Davangere	37.78	43.62	46.35
Total Command	37.86	46.85	48.71

Water Managment

Analysis of satellite data has also helped in evaluating improvement under NWMP in water management aspects such as depth of water applied and sater use eficiency for paddy production.

The depth of water application has improved to 0.859 meters in 1993-94 rabi season over the area commanded by Bhadra Right Bank Canal (RBC) compared to 1.059 meters in 1986-87 rabi rabi season (Table-4) . The depth of water applied on paddy area is 1.018 meter in the 1993-94 rabi season against 1.343 meter prior to NWMP. Similarly water use efficiency for paddy crop improved to 0.478 kg/cu.m in 1993-94 rabi season against 1.343 meter prior to NWMP. Similarly water use efficiency for paddy crop improved to 0.478 rabi season.

TABLE 4. IRRIGATION WATER MANAGEMENT IN BHADRA COMMAND

Rabi season	1986-87	1989-90	1992-93	1993-94
Total Depth of water application (mts)	1.058	1.040	0.799	0.859
Depth applied for paddy (mts)	1.343	1.362	0.946	1.018
Paddy WUE (Kg/cu.m)	0.282	0.396	0.495	0.478

Equity in Water Distribution

The equity in water distribution during 1993-94 rabi season is estimated by the Christiansen's Uniformity Coefficient (UCC) defined as

$$UCC= 1-S(|x_i-x|)a_i/(Sa_i)\ x$$

Where x/x i= depth of irrigation water delivered to incremental area a_i d\weighted mean depth of irrigation water delivered

The UCC value closer to 1 indicates better uniformity in water distribution.

Water supply among subdivisions during 1993-94 rabi seasn is more quitable in Malebennur and Davangere divisions compared to badravathi division (Table 5) . The equity in Water supply among the distributaries within a subdivision is given. Subdivisions with low uniformity are given in bold. There is a need to improve and ensure equitable canal diliverty jpropotionate to the area, starting from these distributaries.

Table 5 Equity In Water Suply During 1993-94 Rabi Season

Christiansen's Uniformity Coefficient among					
Canal	Sub Division	SD1	SD2	SD3	SD4
Bhadravathi	0.752	0.909	0.799	0.750	--
Malebennur	0.929	0.852	-	0.688	--
Davangere	0.919	0.663	0.839	0.934	0.882

*UCC = 1 indicates 100% uniformity in water application

Diagnostic Analysis

Spatial and temporal variability in irrigated area, and crop productivity derived from satellite data provides the capability to diagnose problem distributaries, with low irrigation intensity (Table 6) and low water use efficiency for paddy production (Table 7).

Table 6 Distributaries With Low Irrigation Intensity In Rabi 1993-94

Division	Irigation Intensity (%)		
	90-80	80-60	<60
Bhadravathi	1,21	5,17,22,24,25	-
Malebennur	7, 9b, 10, 13,	12,16	14, 15
Davangere	11b, 11e 3, CHM	SHM, 13, 9c	13,12,12b, HPD

Table 7. Distributaries with low water use efficiency (less than 90% of branch canal average)

Branch canal	Rabi 1993-94	Rabi 1992-93
Bhadravathi	17,22, 25	6,7,8,22
Malebennur	HSD,	HSD, 2 12, 15,16,17
Davangere	1,2,12,11e,15 3,8a, CHM, 9c HPD2	3, CHM, 12a, 12b, 14, 9b, HPD2

Equity within distributary command has been evaluated in reardd to gap in yield between head reach area (HRA) and tail reach area (TRA). Problem distributaries with significant inequity have been identified (Table 8) in Malebennur and Davngere canal divisions.

TABLE 8 DISTRIBUTARIES WITH NON-UNIFORMITY IN PADDY YIELD

Division	Paddy Yield difference between RA & TRA IN Comparison to distributary mean(%)		
	10-20	20-30	>30
Davangere	1,5,6a,13,9c	2,3,9b	12
Malebennur	4,7	-	14,15,LBC, RBC

Such diagnostic analysis and identification of problem pockets wihtin the command area, made possible by satellite data, should lead to corrective management in the subsequent season.

CONCLUSIONS

Multidate satellite data during the rabi season of pre and post-NWMP years have indicated significant increase in irrigation intensity and also in area under paddy crop. since canal discharge is the limiting factor in water delivery rather than reservoir storage, water management is seen to have substantially improved. This is all the more notewoethy since the yield of major rabi crop, paddy, has significantly increased. The depth of water application has decreased under NWMP and water use efficiency for paddy production has improved. Thus there is significant improvement in system performance after NWMP implementation.

Diagnostic analysis of irrigated area and paddy productivity at distributary level, has identified problem distributaries of low irrigation intensity and paddy yield and low/no yield improvement in recent years. Distributaries with large yield gaps between head reach area and tail reach area have been targeted. Canal sub divisions/divisions with low equity in water distribution and distributaries with low water use efficiency for paddy production have been identified. corrective management in subsequent seasons will help in overall and equitable improvement of Bhadra Project.

ACKNOWLEDGMENTS

The authors gratefully acknowledge the guidance and support received from Director and Associate Director of National Remote Snesing Agency. Thanks are also due to Mr. M. L. Lath, Commissioner [Water Management], Ministry of Water Resources, Govt. of India, and Mr. A. A. Pai, then Irrigation Engineer and Mr. Bandyopadyay, Irrigation Engineer, World Bank, New Delhi for their unstinted support and constructive discussions which helped to shape the methodology. Support received from Govt. of Karnataka towards project funding and cooperation during field visits/ data collection is gratefully acknowledged. Various colleagues in Water Resources Group and other functional areas of NRSA have significantly contributed to the success of this study. Thanks are due to Mr. Sanjay who extended support in many ways right through the project progress and in the preparation of manuscript. Secretarial support received from Ms. Sudharani and Mr. T. Subba Rao is acknowledged.

DESIGNING DECISION SUPPORT SYSTEMS TO AID IRRIGATION WATER PLANNING AND MANAGEMENT IN COMMAND AREAS

Developing irrigation potential over a command area requires considerable efforts both in terms of time and money. The main crux of the development of the irrigation facilities lies in making it economically, socially and environmentally viable where the returns from the area are optimized.

Optimization of irrigation water requirements requires proper irrigation scheduling based on the crop water requirements of the different crops sown over an area during that particular season. Remote Sensing and GIS can help extensively in this exercise.

Land-use cover and cropping patterns of the command area for all cropping seasons and irrigation water requirements on area basis for all crops sown are the most important information for proper irrigation planning and management. The information on land-use and cropping pattern are extracted with the help of multi-date satellite imageries as well as existing records and stored as different layers in GIS in the form of spatial and non-spatial data. Crop water requirement for different crops is also appended as an information layer alongwith different data layers of water availability during the entire year, crop calendar, irrigation efficiency, land-use efficiency, soil characteristics, proximity of market places, use of agro-chemicals and fertilizers, socio-economic parameters, climatic and weather conditions, ground water to assess possibility of conjunctive uses etc. All these determine and affect the irrigation water requirement for different crops. Integrating above data layers alongwith analytical models can provide a framework for Decision Support Systems (DSS) those can be explicitly designed to support Decision making process for complex irrigation problems.

GIS analysis help in scheduling the crops over all seasons depending upon the availability of water throughout and optimization of cropping pattern and intensity based on soil and climatic characteristics of the area and suitability and acceptability of different crops that ensures maximum returns. Once these are finalized, water delivery system, distribution and minors etc. can be designed accurately.

ESTIMATION OF RUNOFF FOR AGRICULTURAL WATERSHED USING SCS CURVE NUMBER AND GEOGRAPHIC INFORMATION SYSTEM

A watershed is the area covering all the land that contributes runoff water to a common point. It is a natural physiographic or ecological unit composed of interrelated parts and functions. In India, the availability of accurate information on runoff is scarcely available in few selected sites. However, quickening of the watershed management programme for conservation and development of natural resources management has necessitated the runoff information. Advances in computational power and the growing availability of spatial data have made it possible to accurately predict the runoff. The possibility of rapidly combining data of different types in a

Geographic Information System (GIS) has led to significant increase in its use in hydrological applications. The curve number method (SCS, 1972), also known as the hydrologic soil cover complex method, is a versatile and widely used procedure for runoff estimation. This method includes several important properties of the watershed namely soil's permeability, land use and antecedent soil water conditions which are taken into consideration. In the present study, the runoff from SCS (Soil Conservation Services) Curve Number model modified for Indian conditions has been used by using conventional database and GIS for Karso watershed (Hazaribagh, India).

MATERIALS AND METHODS

Study Area

The study area, Karso watershed, which lies within Damodar Barakar catchment is situated between 85° 23' 30" to 85° 28' E longitude and 24° 12' to 24° 18' N latitude with an elevation ranging from 390-650 m above MSL (Mean Sea Level) and extends over a total area of 2793 ha. The watershed receives an average annual rainfall of 1300 mm and more than 80% of the rainfall is received during the monsoon season (June to September). The minimum and maximum temperature varies in the range of 30 C to 420 C.

The mean relative humidity varies from a minimum of 40% in April to a maximum of 85% in the month of July. The overall climate of the area can be classified as sub-humid tropical. The soil is mainly sandy loam type occupying the maximum area with a land slope varying from 0 to 8 %, maximum slope of some hilly parts of the watershed is upto 22% and the soil depth ranges from 0 to 45 cm.

Land use / Land Cover Map

The conventional land use/ land cover map of the watershed was obtained from the Soil conservation department of Damodar valley Corporation, Hazaribagh, Jharkhand State. Boundaries of different land use were digitized in ARC/INFO and the attributes were given. Five land use/ land cover were categorised in the watershed (Table. 1 and Fig 1).

Fig.1 Land use/Land cover present in the watershed

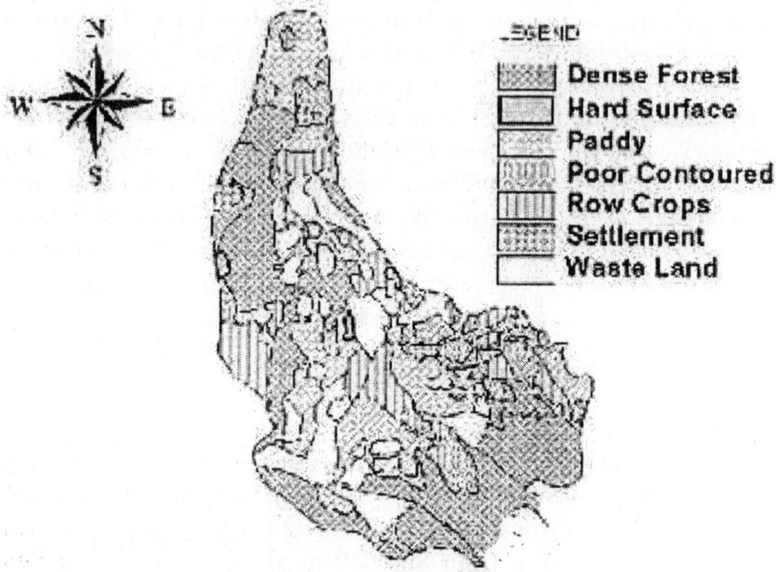

Table: 1. Land use / cover classes present in the study area (Year 1993)

Sl.No.	Land Use/ Land Cover	Area (ha)	Area (%)
1	Agricultural Land		
	Paddy	540.45	19.35
	Poor Contoured	131.35	11.36
	Straight Row	342.85	12.27
2	Settlement	57.25	2.05
3	Dense Forest	970.1	34.74
4	Waste Land (i.e. fallow)	548.50	19.64
5	Hard surface	16.55	0.59
	Total Area	2793.00	

Soils

The soil map of the watershed was traced, scanned and rectified in ERDAS IMAGINE 8.4 by using the registered topographic maps. Boundaries of different soil textures were digitized in ARC/INFO and the polygons representing various soils class were assigned and different colours for recognition (Fig. 2). hydrologic soil groups i.e. A, B, C, and D were considered for the classification of soils of the watershed.

The soils of group A were of low runoff potential, high infiltration rate, the soils of group B were of moderate infiltration rate, moderately well drained to well drained.

The soils of group C were of moderately fine to moderately coarse textures, moderate rate of water transmission and the soils of group D were of slow infiltration and high runoff potential.

Fig.2. Soil Map of the watershed

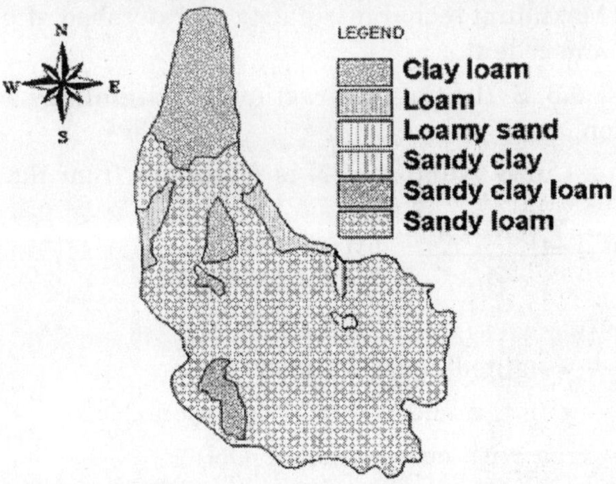

The Curve Number Method

The Curve Number method (SCS, 1972), also known as the Hydrologic Soil Cover Complex Method, is a versatile and widely used procedure for runoff estimation.

In this method, runoff producing capability is expressed by a numerical value varying between 0 - 100. In the past 30 years, the SCS method has been used by a few researchers because it gives consistently usable results (Rao et. al.,1996; Sharma et al. 2001; Chandramohan and Durbude, 2001; Sharma and Kumar, 2002) for runoff estimation. The SCS method with initial abstraction consideration is given below:

The curve number method assumes that

$$S = \frac{25400}{CN} - 254 \quad (1)$$

$$Q = \frac{(P - 0.3S)^2}{(P + 0.7S)} \quad (2)$$

Where,

Q = Runoff depth, mm

P = Rainfall, mm

S = Maximum recharge capacity of watershed after 5 days rainfall antecedent

I_a = 0.3 S (Initial abstraction of rainfall by soil and vegetation, mm)

CN = Curve Number, CN is found out from the table.

$$CN = \frac{\sum (CNi \times A_i)}{A} \quad (3)$$

Where

CN = weighted curve number.

CN_i = curve number from 1 to any no. N.

A_i = area with curve number CN_i

A = the total area of the watershed.

The land use/ land cover map and soil map were intersected in command tools of ARC/Info. The areas of different land use class and soil combinations were obtained in the attributes selection menu by using logical expression and accordingly different CN (Curve-Number) values were assigned. Thus weighted value of CN for the whole watershed was worked out. The CN values for AMC II condition can be converted in to CN values for AMC I and AMC III condition by using available conversion factors as suggested by Suresh, 1997. Putting the value of Curve Number in formula (1), the recharge capacity 'S' was calculated. The direct runoff of the watershed was calculated using formula (2) .

The daily rainfall data for the years 1993 of the study area and the values of maximum potential retention, S obtained from the weighted CN for the watershed area were used for the estimation of runoff from SCS (model). The estimated runoff was then validated by comparing it with the four selected events of monsoon season.

Results and Discussion

Based on the hydrological soil group, the maximum area of Karso watershed was observed to be under hydrological soil group B (84.32%) followed by (15.23 %) C and (0.45 %) group A. Similarly, the study area was identified in the five major land use classes namely agricultural land (42.98%), dense forest (34.74 %), wasteland (19.64 %), settlement (2.05%), and hard surface (0.59%). The major area of the watershed was under agricultural land. The agricultural land was divided into three categories i.e. area under paddy cultivation (19.35%), straight row crops (12.27%) and poor contoured (11.36%) crops. The USDA curve number table 2 (Tripathi, 1999) modified for Indian condition was used for the determination of the curve number for the watershed. By intersecting the land use map and soil map the curve number was assigned to the each combination of land use and soil type. Weighted value of CN were found out to be 71.42 52.75 and 85.7 for AMC II, AMC I and AMC III conditions respectively. The daily rainfall data for the year 1993 were collected and the weighted curve number of the watershed was used for the estimation of runoff. The calculated runoff was found out to be 244.40 mm for monsoon season of the year 1993 which is approximately 35.52 per cent of the total rainfall.

Table: 2. Status of different soil group presents in the study area

Sl. No.	Soil Group	Area (ha)	Area (%)
1	A	13.09	0.47
2	B	2354.72	84.30
3	C	425.19	15.23
		2793.00	

In order to validate the SCS model output, estimated runoff was compared with measured runoff for four selected events of the monsoon season(Table 3). The maximum and minimum deviation were observed to be 28.33 and 3.27 per cent respectively, which are within the permissible limit.

Table: 3. Comparison of estimated runoff with the observed runoff

Storm Date	Rainfal l(mm)	AMC Condition	Runoff Observed (mm)	Estimated Runoff (mm)	Percent deviation
11-07-1993	43.00	II	1.54	1.37	11.03
25-07-1993	21.1	I	17.18	12.33	28.33
21-08-93	105.6	III	85.92	63.78	25.76
13-10-1993	72.2	III	34.37	34.73	1.04

In the present study the methodology for determination of Runoff for Karso watershed using GIS and SCS model was described. This approach could be applied in other Indian watersheds for planning of various conservation measures.

PEDOGENESIS IN RELATION TO PHYSIOGRAPHY IN SOILS

Morphological, physical and chemical characteristics of soils developed on different physiographic units in Ropar Kandi tract in Punjab State (Northwest India) revealed two distinct stages of pedogenic development : A-C and A-(Bw) - C profiles. The young and strafied soils from unstable landforms (hills, valley, piedmont plain, recent and active flood plan) representing newer alluvium are classified as Typic Ustipsamments, Typic Ustorthents and Typic Ustifluvents. The soils develoed on relatively stable landforms (old flood plain) show development of cambic or structural B horizon and are classified as Typic variation and thus, is separately described both for 'East and West' of Satluj River.

The differences in soils developed on various physiographic units are primarily due to the nature of parent material, topography and to some extent age or time. The study established a well-defined relationship between physiography and development of soils.

An intimate knowledge of soil resources of a region is necessary for planning any agricultural development. In this context, remote sensing have opened new vistas in soil resources surveys. The knowledge of physiographic processes as portrayed on satellite imagery in the form of surfacial features provide enough traits to delineate soil boundaries. This technology has been employed successfully for soil mapping in different parts of the World (Westin and Freeze, 1976: Gastellu- Etchegorry et al., 1990; Abd-El-Hay et. al., 1991) and in India (Sehgal et. al., 1988; Rao et. al., 1989; Kudrat et. al., 1992). Information about the pedogenesis of soils in relation to physiography in Ropar Kandi tract is lacking. With this in view, the present study was undertaken to evaluate the contribution of different soil forming factors to the development of soils from different physiographic units in the area.

GENERAL DESCRIPTION OF THE AREA

The study area, a part of Kandi tract in Ropar district Punjab lies between 30°50' and 31°13'N latitudes and 76°20' and 76°39' E longitudes. It has distinct physical agro-ecological characteristics which are different from those in other parts of the State. The area was undulating topography, inadequate ground water, steep slopes, bare land surfaces, lack of vegetation due to deforestation and overgrazing and problem of soil erosion. It is located both on the west and east of river Satluj, towards upstream of Ropar Headworks up to Anandpur Sahib and covers an area of 47,502 hectares. The area is bounded by Himachal Pradesh in the east and district Hoshiarpur (Punjab) in the west. Geologically, the area consists of para autochthonous upper Siwalik belt, piedmont plain and alluvial tract. The climate is the semi-arid (sub moist) and the area has ustic soil moisture regime and hyperthermic soil temperature regime.

Materials and Methods

The IRS-1A LISS-II data in the form of false colour composites of November 1991 generated from bands 2,3 and 4 was visually interpreted for delineating various physiographic units viz., hills, valley, piedmont plain and flood plain. Each unit was further sub divided based on combination of tine, texture, land type, texture, pattern etc. These physiographic units have been further sub divided based on land parameters like slope, erosion, natural vegetation, drainage and landuse, which have weightage on the soil variation and land management. The study area is quite heterogeneous with respect to soil variation and thus, is separately described for both 'East of Satluj' and 'West of Satluj'. The pedons representing major soils were studied for their morphology and physico chemical properties as per standard procedures. The soils were classified as per Soil Taxonomy (Soil Survey Staff, 1975) and mapped as association of soil families. The relationship between physiography and soils was established. The final physiography - soil map was prepared on 1:50,000 scale.

RESULTS AND DISCUSSION

Physiography and Soils:

Systematic approach of image interpretation along with sufficient collateral information regarding geology, climate etc. of the area revealed the following physiographic units:

1. Hills: This unit is represented by Upper Siwalik belt which consists of massive, poorly lithified sandstone and silt stone. These consists of beds of silt and clay with rounded pebbles and gravels. The hills are covered with sparse vegetation, bushes and shrubs. It has moderate to very slopes and is severely eroded. This unit is further sub divided into
   (i) Hill side slopes and
   (ii) Hill foot slopes.
2. Valley: The valleys are narrow to broad with gentle to moderate slopes. The ground water is mostly at shallow depth. The valleys in the area are found in different location and has been subdivided into (i)Hill valley (ii) River bed in plains.
3. Piedmont plain: It represents plain with a gently rolling slope at the base of the hills, intercepted and dissected by seasonal rivulets. This units is partly cultivated and partly under bush vegetation. The deposits near the hills are finer - in texture, in general. This unit is

divided into (i) Upper piedmont plain (ii) Lower piedmont plain. Further sub division is done on the basis of landuse, erosion, slope and soils. The piedmont plain 'West of Satluj' consists of coarser material and differ from the deposits of severely dissected piedmont plain 'East of Satluj' which consists of beds and silt and clay along with gravels. A small pocket of piedmont plain south of Siersa River shows mixed characteristics.

4. Flood plain: The flood plain is the result of deposition due to flooding of river Satluj and river Sirsa and it occurs along these rivers forming narrow strips. It is characterized by low relief and consists of pebbles, gravels in sandy and silt matrix. The deposits away from the river are finer as compared to those in the vicinity of the rivers. The eastern side flood plain exhibit comparatively finer material than its western counterpart. The flood plain is divided further according to its chronosequence into (i) Old flood plain (ii) Recent flood plain and (iii) Active flood plain. These units are further classified based on landuse and variation in soils. The general characteristics and soil physiography relationships has been given in Table 1 and 2 separately both for 'East of Satluj' and 'West of Satluj'.

MORPHOLOGICAL AND PHYSICO CHEMICAL CHARACTERISTICS

The soils representing different physiographic units were studied for their morphological and physico-chemical characteristics. The morphological characteristics of the soils reveal that they have 10YR hue except for a few horizons in imperfectly drained soils of the active flood plain which have 2.5 YR hue colors. The values range from 3 to 6 and chromas 2 to 6.

The soils of hill foot slopes, upper piedmont plain (West of Satluj) and active flood plain are sandy, containing more than 85% sand. These soils are very low in organic matter, silt and clay content and thus lack structural peds, being single grain in all layers. The soils of lower piedmont plain lack profile development. The old flood plain soils show weak fine to medium moderate sub angular blocky structure in the B horizon.

The soils fo the area show normal pH, low organic carbon (0.01 to 0.4%) and low electrical conductivity (0.11 to 0.36 dSm-1). The CaCO3 content is low (0.1 to 4.8%).

The flood plains soils show irregular distribution of $CaCO_3$ with depth, reflecting parent material stratification. The cation exchange capacity (CEC) and extractable bases vary considerably among the studied soils. The fine textured soils from old flood plain have higher CEC compared to the coarse textured soils in other units.

SOIL CLASSIFICATION

The soils of the area have been classified according to the criteria proposed by Soil Taxonomy (Soil Survey Staff, 1975) based on morphological and physico chemical characteristics.

The soils have been classified into 2 orders i.e. Entisols and Inceptisols.

1. Entisols: These coarse textured soils are young, show no diagnostic horizon except for ochric epipedon. There is irregular distribution of organic carbon with depth. These soils have received new sediments as fast or faster than the assimilation of previous material into the genetic horizons and as such these soils are highly stratified. In these soils, either the operation of pedogenic processes is obstructed or lack of time is the limiting factor of pedogenic evolution of diagnostic horizons. At sub group level these soils qualify for "Typic' sub group; At places, these qualify for 'Aquic' to 'Anthraquic' sub groups as they show characteristic associated with witness.
2. Inceptisols: These moderately developed soils of piedmont plain, old and recent flood plain show the presence of altered textural or structural B horizon i.e. Cambic horizon (Bw). They show moderate pedogenic development. At sub group level these soils qualify for 'Typic' sub group.

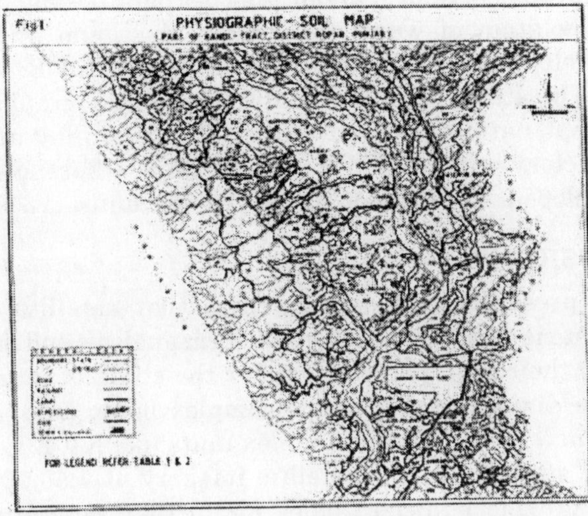

Influence of Soil Forming Factors

The morphological and physico chemical characteristics of soils from different physiographic units revealed two distinct stages of pedogenic development: A-C and A-(Bw)-C. The young and stratified soils from unstable landforms (bills, valley, piedmont plain, recent and active flood plain) without any diagnostic horizon except ochric epipedon, showed little alteration of the original deposits. Topography of the land along with the time factor and nature of parent material seemed to have played a vital role in the genesis of these soils. The areas in the vicinity of the Satluj and the seasonal rivulets receive coarser sediments from hills and piedmont plain because of the steep gradient from the Siwalik hills to the Satluj river.

The soils of the Siwalik hills experience severe erosion, which prevents the maturation of soil profile. The soils of the flood plain (recent and active) are composed of coarser sediments. Because these soils are frequently flooded during the monsoon season, severe erosion and reworking of older sediments and deposition of fresh ones leave little time for consolidation of sediments into aggregates. These soils were deposited by the meandering action of Satluj river, as evidenced by clear stratification and absence of diagnostic horizons except for ochric epipedon. The soils from relatively stable landforms (old flood plain) on nearly level to gentle slopes permit comparatively more percolation of water throughout the soum and hence show development of a colour and/or structural B (cambic) horizon. The study of soils from different physiographic units reveals that nature of parent material, topography and time are the factors responsible for the pedogenic differences in the soils developed on different physiographic units.

CONCLUSION

The present study demonstrates that satellite remote sensing data is a valuable tool for landform analysis and mapping of soils in their total perspective. For the study of soils which is a three dimensional dynamic complex body, field work is always a necessity. In some complex units like K1, K3, F1 and A1 in the study area, the satellite imagery at 1:50,000 scale does not provide sufficient clues for further separation and mapping of soils. For that detailed investigations are needed at scale larger than 1:50,000 supplemented with aerial photographs.

4

ASSESSING BIOLOGICALLY DEGRADED SOIL

It is practically impossible for man to produce stable humus synthetically. Man can only properly cultivate the field, supply organic matter and encourage the development of stable humus in the soil. Soil with stable humus must always be protected to maintain the fertility and productivity. Stable humus is the crucial center, the focus of the soil life cycle. Adhering to farming practices that assure the production of stable humus becomes the farmer's main objective. In this paper, the authors discuss the various steps involved in assessing the biological state of the soil, by calculating humus content.

INTRODUCTION

For nearly one hundred years soil science in most schools of higher learning especially in agricultural sciences has been primarily concerned with the physical and mechanical aspects of soil structure. Biological thinking or biologically degraded soil has become a major concern only in last few years. The new approach consider not only the physical properties and mineral structure of the soil, but also the process by which organic matter is transformed into humus by microorganisms. Good soil consists of 90% mineral and 10% bio-organic substances. The bio-organic parts consists of 70-90% humus, 10-30% roots, and active fraction, which constitutes living organisms and other, constitutes.

Some of the nutrients get lost through leaching due to continually wet weather, melting snow, flooding or through denitrification. Also each cultivated plant takes away nutrients from the soil, as soon as the crops are harvested. The substantial task for the farmer is to take care in returning nutrients taken from the soil through harvesting. The conventional farmer is uses water soluble, mostly salty chemical fertilizer. In contrast, the bio-organic farmer uses organic matter in the form of crop residues, and other wastes, and/or compost, in order to take of soil life, its proliferation and simulation to highest effectively. Microbial soil compound plays one of the most important roles in the decomposition of organic litter. Soil organic matter includes plant and animal remains in various stages of decomposition, cells and tissues of soil organisms, and substances produced by the soil microbes. Well- decomposed organic matter forms humus-a dark brown, porous, spongy material with a pleasant, earthy smell. The occurrence of microorganisms in the soil depends on many factors e.g. on soil acidity, air and soil humidity, air and soil temperature, soil water, all meteorological factors etc. Seeing an importance of soil biology this study only considers the mineralization of organic matter without erosion. A distinction is made between non-resistant organic matter (i.e. fresh organic residues) and resistant organic matter (humus). The biological degradation dealt in this study is as mineralization of resistant organic matter mainly based and soil parameter, which calculates the decay rate of humus.

STUDY AREA

The chosen study are in this study is India, lies to the north of equator, between 8o4' and 37o6' North and 68o7' and 97o25'. East It is bounded in the south by the Indian Ocean, in the west by the Arabian Sea, in the east by Bay of Bengal, in the north-east, north and a part of the north-west by Himalayan ranges, and the rest of the north-west by the Great Indian Desert. Physiographically, India can be divided into seven regions: the Northern Mountains, constituting the Himalayan ranges; the great Indo-Ganagetic Plain; the Central Highlands; the peninsular Plateau in the states of maharashtra, Andhara pradesh, Karnataka and Tamil nadu; the East Coast; the West Coast; and the Island. The average elevation of the plateau that dominates most of Peninsular India ranges between

300 and 1800 m. The river systems of India may be broadly classified into two groups: the Himalayana and the Deccan. The main rivers of the Himalayan group, which are snow-fed, are the Indus, Ganga and Brahmaputra including their major of the country. The Deccan system include the west-flowing Narmada, Tapi and Periyar and the east-flowing mahandi, Godavari, Krishna, Penneru and kaveri.

The Republic of India's mainland stretches 3214 kms from north to south and 2933 kms from west to east. The total land area is 3.29 million sq. kms. Size wise it is the sixth largest country in the world. India exhibits a variety of landscapes and climatic conditions in the evolution of different soils and vegetation. It shows a string relationship among the soils, climate and vegetation.

METHODOLOGY

Calculating Soil Organic Matter/Humus Decay Rate

In general, the organic matter in the soil comes from plant and animal residues when organic matter decomposes, by the action of microorganisms and other forces. It produces substance and nutrient that helps to support plant life. Well -decomposed organic mater is called humus, which helps to hold moisture in the soil. In this study humification and mineralization are the two different combined processed used for quantifying the soil organic matter/humus decay in a year.

What is Humification?

Humification, the transformation of organic matter into humus leads to soil decay, is a fascinating process. Organic materials such as manure or field wastes, when disked into the upper six inches of topsoil will undergo several changes. The humification process involves first catabolism, then anabolism. These are not truly correct terms as they are usually used for same functions within living organisms, but we may soil as one living organism.

What is Mineralization?

Is the process of freeing minerals from organic molecules (carbon bonds). During humification there are two possible end products for atoms within the starting molecules. Minerals may either build up stable humus or be free in form; carbon either tied within humus or evolves in form of carbon dioxide. The following is a summary table of where atoms end up after humification.

All minerals within organic compounds	80% freed up and 20% in humus
Carbon from carbohydrates	80% evolves as carbon dioxide and 20% goes to humus formation.
Nitrogen	Upto 50% will remain in humic form.

RESULTS AND DISCUSSION

Figure. Annual Decay Rate of Humas in India

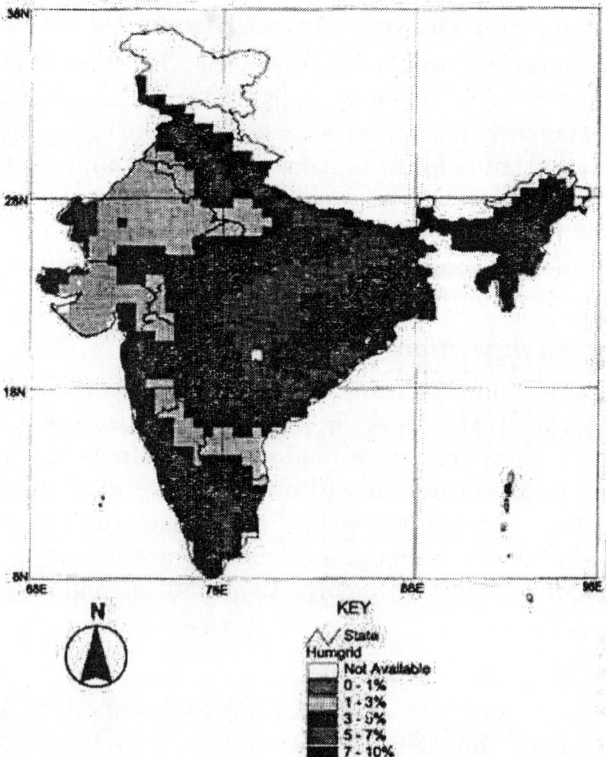

The result of calculating the humus decay rate by Koef equation is classified on the basis of their severity and value ranges. Six classes have been used. Following ratings were adopted in relation to the climatic index as shown in table 1 below. The rate of humus decay in percentage is mapped in Figure below.

Table 1. Humus decay rate on the basis of climatic index

Class	None to Very Slight	Slight to Moderate	Moderate to High	High to Very high	Very High to Severe	
Rating(%/yr.)	0-1%	1-3%	3-5%	5-7%	7-10%	>10%

Figure 3: Effect of Soil Parameter Over Soil Orgainc Matter

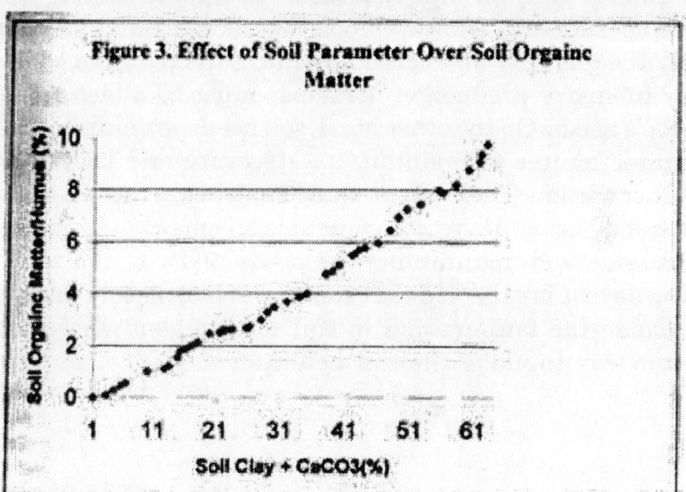

Soil organic matter plays a central role in forming soil aggregates. Soils with high aggregate stability are more resistant to degradation by erosion and compaction. As the clay content of a soil increases, it requires a higher content of organic matter to maintain a given level of aggregate formation and stability. In this study it was found that, increase in clay and $CaCO_3$ content of soil correspondingly shows possession of high organic matter as shown in figure 3. Soil organic matter acts as glue due to its polysaccharide content, holding soil particles in stable aggregates.

The rate of decay varies according to the texture of the soil, being faster in a sandy than in a clay soil, the nature of the organic matter, the pH and water content above field capacity. Soil reaction between pH 5.0 and 7.5 has less effect on biological degradation. Slope aspect influences soil temperature and humidity, but slope angle has little influence on biological degradation. Altitude is taken into account by the climatic index, through its effect upon temperature.

CONCLUSION

The optimum organic matter content of soil depends on local climate, the amount and type of clay material present in the soil, and the soil, and the soil's intended use. Ideally, in a good-quality soil, all the functions of organic matter listed earlier operate fully. However, soil organic matter alone cannot supply the quantity and balance of nutrients required by crops under intensive production; fertilizer must be added to meet the crop's needs. On the other hand, soil needs optimum amounts of organic matter to maintain its structure and keep it in a tillable condition. The current study shows that the soil organic matter and humus decay rate content is an important component in assessing and maintaining the productivity of the soils. In coming days it is expected that concern of biologically degraded soil among the farmers and in soil science as a whole might a better way to make the soil healthier.

5

Research on Mobile Mapping System and its Application in Precision Agriculture

INTRODUCTION

Mobile Mapping System is an information technology that has been developed since late 1980s with the advance and progress in mobile positioning technology, modern communication technology, spatial information technology and embedded technology, etc. Integrating diversiform advanced technologies above-mentioned, Mobile Mapping System is capable of real time data accessing, mapping,and spatial data visualization. Mobile Mapping System not only collects dynamic data about some object in moving, but also manipulates these data in order to make decision efficiently, and make these data be downloaded into a desktop GIS simultaneously by wireless or wire communication as you like. Now Mobile Mapping System has been applied in many fields, such as intelligent transport, precision agriculture, field surveying and environment engineering and so on for outstanding real time supply is very essential in these fields and Mobile Mapping System can do.

Precision agriculture (precision farming, digital agriculture) has evolved from a concept ten years ago into an emerging technology today. Its basic purpose is to manage farmland by the square meter instead of the square mile. It is certain that precision agriculture can be described as the next great evolution in agriculture. With Mobile Mapping System, every inch soil can be monitor and planting data in detail can be acquired anywhere anytime. How to use Mobile Mapping System in Precision Agriculture will be also discussed in this paper,

MOBILE MAPPING SYSTEM CONFIGURATION

Mobile Mapping System consists generally of three parts, that is moving platform, navigation sensors and mapping sensors. Its configuration can be described as follow (figure 1).

Figure 1 Mobile Mapping System Configuration

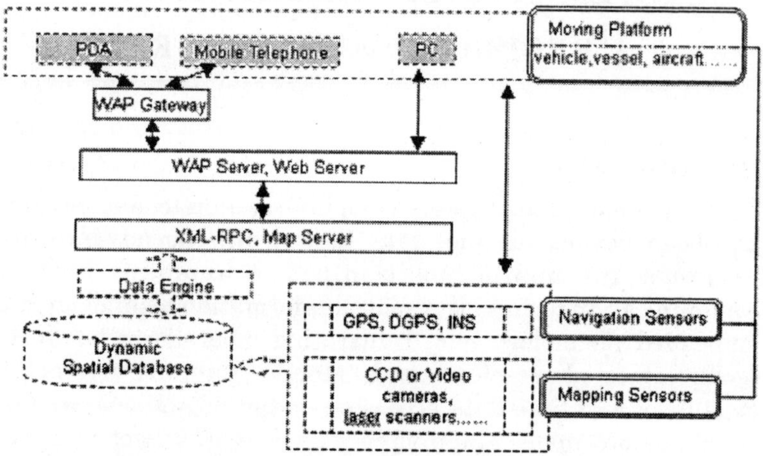

The mobile platform is designed to equip data acquisition devices including mapping sensors and positioning apparatus. It may be a land vehicle, a vessel, or an aircraft. Thus the system can observe objects at closer range flexibly. Generally, Mobile Mapping System monitor objects by CCD cameras, video cameras, laser scanners and radar sensors.

These mapping sensors are selected according by purpose, but they must obtain high flexibility in data acquisition, more information with less time and effort, and high productivity.

Of course GPS (Global Positioning System) must be connected at the same time in order to record the position and time. Mobile Mapping System is essentially useless without GPS. GPS is developed and initiated by the United States of America. The system has capability of navigation and positioning anywhere on earth, anytime, and under any conditions. Since SA (Selective Availability) was permanently turned off at midnight May 1, 2000, the accuracy of GPS data has been increase. And differential GPS techniques can help to obtain high positional accuracies. In addition, vehicle wheel sensors and INS (Inertial Navigation System) are important navigation instruments, too. Integrating with GPS, they can provide both the track of the vehicle and position and orientation information of the mapping sensors.

All these data acquisition devices can get plenty useful attribute information. The information will be edited and geo-referenced, then stored in the Spatial Database. Geo-reference process is very necessary because these data will be integrated with other information from multi-sources.

Field data collection is always a difficulty for cartographers, surveyors and researchers. The tools available for mapping applications have been bulky in size and weight, expensive, and difficult to learn for a long time.

Fortunately, the advances of remote sensing, GPS technology, GIS and some data edit and analysis software drive the field data collection. The advance refers not only precision has been improved, but also the hardware has become smaller, lighter, and cheaper. And the software has become easier to learn, and more inexpensive. So the data collection task becomes easier, more economical and faster to complete.

Mobile terminals should be another important component of Mobile Mapping System. Spatial data can be displayed, edited, or analyzed with them. There are different terminals that can finish such task. They are PDA, Mobile telephones, and Land Phones. These handheld equipments can exchange data with desktop PC via serial or USB ports, and some can do this by Internet access. In many cases, Mobile Mapping System is equipped on vehicle, and above functions can be done on PC. Whatever terminals can interchange data with GPS data, and access data from long-distance database by Internet or wireless communication equipment. So the works need the support of modern communication technologies.

Key Technologies

Integrates space information, modern communication and intelligent technologies, Mobile Mapping System is designed to collect real time field data with certain positioning information, editing, processing, visualizing, and mapping multi-source data. In its working, following key technologies exert important functions.

Intelligent Integration of Multi-Dimension Data

Usually field data derived by Mobile Mapping System at scales have to be combined with ancillary data layers from other sources. The resolution and format of multi-sources should be matching. Moreover it is necessary for GPS data properly matching with the geo-spatial database. Recently the integration method has been much more intelligent for dealing with and integrating quickly and accurately. Such study need further progress though great deal of work has been done on it.

Compression and Decompression Models of Spatial Information

Compression and decompression models study has been discussed for a long time. Almost every spatial database involves the problem because large data files not only take much more capacity of the database, but also at the same time its loading cost much longer time than small files.

Now there are many compression and decompression models for spatial information. All of them have their own characteristics and their compressibility is different from each other. Generally higher compression ratio brings more information loss. When selecting the compression and decompression models, the safety, time cost and efficiency are main factors considered.

Efficient Transmission of Spatial Information

Spatial data/information includes not only remote sensing data, survey data, but also graphic data and attribute data, so transmission of the large data is a difficulty especially when telecommuting or data synchronization between mobile terminals and desktop GIS. Thereby efficient transmission of spatial information is needed imminently. To resolve the problem, bandwidth of the broadband networks should be improved. At present the new networks with 10 G/s transmission speed is building in USA. It is obvious that the transfers of large data depends on broadband networks technology.

Interface of Wireless Communication and Mobile Intelligent Terminal

The capacity and computing ability of mobile intelligent terminal are limited, so it is impossible to store all the data in the limited memory and flash card. Thereby it is absolutely necessarily to transfer the data by Wireless Application Protocol (WAP) from server to user. Accordingly it becomes a key problem to make the interface friendly, and at the same time the data reading speed is another issue settled imminently. Thereby the study on interface of wireless communication and mobile terminals has been paid on much more attention.

APPLICATION IN PRECISION AGRICULTURE

Precision Agriculture

Precision agriculture is a new agriculture technology system development quickly in recent years. Integrating agronomy, geography, biology, agrology, botany, geo-spatial science, precision agriculture can be defined as a comprehensive system designed to help farming. It involves various problems in crop planning and includes tillage, planting, chemical applications, harvesting, and post harvest processing of the crop. Precision farming is a pro-active approach that reduces some of the risk and variables common to agriculture, so we think that it has the potential of optimizing cost and ecological effects through the application of crop information, advanced technology and management practices. Accordingly, how to get these information and data quickly and accurately becomes the foundation of precision agriculture construction, while how to manage the information and make decision quickly and intelligently is the key technology in precision agriculture.

It is obvious that information technologies are foundation of precision agriculture, and positioning, timing, mapping and analysis are most important among them. Accordingly Mobile Mapping System can help agriculturist with a new capability of gathering information for implementing decision-based Precision Agriculture.

Mobile Mapping System in Precision Agriculture

For precision farming, information technologies are so essential that Mobile Mapping System is important or beneficial. Equipped with mapping sensors and navigation sensors, Mobile Mapping System can collect field data anytime, anywhere, in any manner. At the same time, Mobile Mapping System can connected with a GIS or combine GIS software in its mobile terminals directly. Thus Mobile Mapping System can monitor planting process all along. Moreover it can contrast multi-temporal data collected or stored in database and find what changes occurred, where, when and how, then give an efficient plant plan. What function the Mobile Mapping System has in precision agriculture is narrated in Figure 2.

Figure 2 Function of Mobile Mapping System in precision agriculture

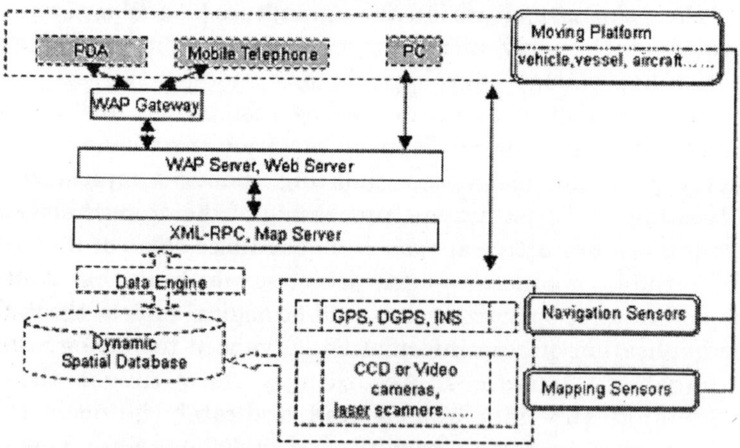

Improve the Accuracy of Soil Sampling

Soil conditions and soil quality influence crop growth greatly. Usually how often do farmers fertilize the soil or water the land depends on the soil conditions and plant type. So it is a basic work to master the soil conditions. From the view of precision agriculture, soil is different meter by meter. Thereby soil sampling becomes the foundation step. The accuracy of soil sampling is requested higher.

Accuracy of soil sampling refers mainly to accurate degree of the position information where the soil samples were taken. With the accurate knowledge about the coordinate location of the soil samples, a soil data layer can be developed accordingly. With the accurate position information, navigating back to those locations for re-sampling is possible.

Mobile Mapping System has equipped navigation sensors, such as GPS and INS, so the time and position information can be recorded at the same time when soil samples are taken. Consequently a soil difference map can be create, on which physical attribute of soil is described.

Plant Growth, Diseases and Insect Pests Monitor

Generally precision agriculture is constructed in a large area. Hereby it is impossible for all plants growing in like manner. Sometime only a single part suffers the diseases and insect pests. Then it is unnecessary and even harmful for all plants to spray pesticide or other medicine in the same way. Different solution should be taken according to the real conditions of plants.

Mobile Mapping System can finish the monitor task of all the plants. Using the scanners, the peculiar plants can be found quickly, and its position also can be recorded. In fact satellite remote sensing data also can provide such distinction information. Compare with Mobile Mapping System, its spatial range is much larger but space resolution is lower and constrained by satellite calendar.

Analysis of the Crop and Field Information

The purpose to get so much data about the soil, plant and so on is to master the planting conditions and to make decisions for all planting process. Accordingly GIS and Agriculture Expert System are usually imported in precision agriculture in order to edit, process, integrate and analyze the crop and field information and get corresponding resolve scheme.

Mobile Mapping System may Load multi-source information and different types of agricultural data to master the conditions of the field. Comparing the information or data, relationships within and between data sets can be found. That is the relationships among different factors can be made certain. By the relation, agronomist or Agriculture Expert System can make production plan.

After the production plan has been made, farmers can command the farming machines to work automatically, for these machines has been equipped intelligent implement and positioning devices. Moreover the entire process of one year farming also can be recorded and evaluated, and the experience can be analyzed to help next year work.

Discussion

Mobile Mapping System is never a concept. Many applied Mobile Mapping Systems have been studying in many counties, and some have been used in practice. The Photogrammetric lab of the Institute of Geomatics at Swiss Federal Institute of Technology has developed a helicopter based portable handheld MMS for avalanche mapping. The Institute of Remote Sensing Applications of Chinese Academy of Sciences has developed a multi-sensor mapping system called Airborne Laser-Ranging and Multi-Spectral Imaging Mapping System (ALMIMS). It is clear that Mobile Mapping System can be used in many fields. There are many such samples. Its application in precision agriculture can be identified as available and efficient in this paper.

Mobile Mapping System technological development speed is still very rapid reasonably. With the development of spatial information technology and modern communication, as well as the advance in computer hardware and software technologies, Mobile Mapping System will explore new areas of applications and become more intelligent, efficient and steady. It may run online GIS module for an unmanned aerial vehicle, motion innovative intelligent traffic management and service, etc.

6

RAINFED FARMING DEVELOPMENT

The Rainfed agriculture plays an important role in Indian economy. In India 68 percent of total net sown area (136.8m.ha) comes under rainfed lands spread over 177 districts. Rainfed crops account for 48 percent area under food crops and 68 percent of the area under non-food crops. Nearly 50 percent of the total rural workforce and 60 percent of livestock in the country are concentrated in the dry districts. In Tamil Nadu, out of 7 m.ha of cultivable area, around 3.1m.ha comes under rainfed agriculture. But uncertainty in production due to fluctuations in total rainfall and changes in its distribution, decrease in relative productivity in rainfed lands etc. affect the livelihoods of many poor and marginalised farmers. Considering the importance and issues of rainfed farming, DHAN Foundation initiated an experimental project on Rainfed Agriculture with following objectives.

- Understanding the status and issues of rainfed farming in the locations selected and identification of specific interventions.
- Improving the total factor productivity of rainfed farming.

The new theme was launched on October 2, 2002 on the occasion of the DHAN Foundation Day celebrations. The pilot project is in the initial phase of implementation. In order to understand rainfed farming context in South India a farmers workshop was organised on the theme *'Rainfed Farming – Past, Present and Future'*. Farmers and experts in this field attended the workshop. The focus of the workshop was on understanding the changes in rainfed farming in the last twenty years and issues faced by these farmers. The prevailing best practices in rainfed farming were also shared. The deliberations in the workshop were on generating ways and means to address the issues that had been identified.

After thorough consultation and screening, it was decided to experiment on this theme in two different agro climatic zones, one in Kallikudi block of Madurai district, which represents black soil. The second one is Natrampalli block of Vellore district, which represents red soil and hard rock areas. In this block dependence on groundwater is present in part of it, while the rest is predominantly drylands. The following areas have been taken up for experimentation in the rainfed farming development theme.

- Understanding and evolving suitable community organisation model
- Transfer of farmers best practices for wider use
- Identification and transfer of suitable technologies evolved by the research institutions
- Finding suitable extension processes and arrangements for the same
- Finding solutions to issues not addressed by best practices and modern technologies.

Focus Areas

1. Creating a community organisation base for the theme in the new location by designing a suitable Community Organisation model for rejuvenating rain fed farming
2. Understanding the rain fed farming practices in the location.
3. Trying out few experiments to know their suitability to the location.
4. Placing a team & building their capacity on the theme.

Progress

The rainfed farming development theme has evolved a people organisation model for piloting. Uzhavar kuzhu (Farmers groups) have been formed with 15 to 20 farmers who have lands at close proximity as members. 126 Uzhavar kuzhus have been promoted with 2039 members with both men and women. Kalanjiams are formed with landless agriculture labourers and for tenant farmers. 16 Kalanjiams have been promoted with 249 members. The team has promoted twelve watershed associations.

Situation specific and need-based interventions were attempted on experimental basis. These attempts were centred on the local best practices and indigenous technology. Focus was given on individual farm that too with small and marginal farm holdings; where the vulnerability is very high. Landless farmers are given central focus, providing them access to lands is one of the key purposes of the theme. The team has evolved several areas for observation and experimentation. The resources for the work have been mobilised through the NWDPRA (National Watershed Development Project for Rainfed Areas) programme. Observation studies have been taken up with 123 farming families. Experimentation on technologies suggested by the agricultural universities and the indigenous ones were attempted to find the suitability for the selected locations. The areas in which experiments were taken up are improved agronomic practices, plant protection measure, crop improvement and farm mechanization. The farmers were trained to observe the performance of each experiment. Farmers could see positive results in some of the experiments. They are interested to follow them in the coming seasons.

The team has built good relationships with District Nodal Agency (Department of Agriculture).

The team could undertake entry point activities like deepening of drinking water wells and construction of hand operated bore wells to solve the drinking water problems in the villages in collaboration with them. Drought proofing works were undertaken with the grant support from NOVIB, the Netherlands in both the locations. The teams gained a lot of learning which was very useful to define the intervention areas for the theme.

An Advisory board has been constituted with eminent persons with experience in this field at time of initiation of the theme. The advisory board met once and the theme related progress including the observation and the experimentation studies were shared with them to get their feedback and suggestions. Several capacity building events were organised for the team members to build their knowledge and skills on community mobilisation and technical aspects.

The team undertook and completed two research studies namely a) Factors responsible for cropping pattern change in Thirumangalam Taluk of Madurai district and b) An analysis of conversion of dryland into garden land in Nattarampalli block of Vellore district . These two studies were useful in understanding the context of the locations in more detail.

AGRICULTURE IN INDIA: ISSUES AND CHALLENGES

"Agriculture is the backbone of the Indian Economy"- said Mahatma Gandhi five decades ago. Even today, as we enter the new millennium, the situation is still the same, with almost the entire economy being sustained by agriculture, which is the mainstay of the villages. Not only the economy, but also every one of us looks up to agriculture for our sustenance too.

Significance of Agriculture

Although agriculture contributes only 21% of India's GDP, its importance in the country's economic, social, and political fabric goes well beyond this indicator. The rural areas are still home to some 72 percent of the India's 1.1 billion people, a large number of whom are poor. Most of the rural poor depend on rain-fed agriculture and fragile forests for their livelihoods.

The sharp rise in foodgrain production during India's Green Revolution of the 1970s enabled the country to achieve self-sufficiency in foodgrains and stave off the threat of famine. Agricultural intensification in the 1970s to 1980s saw an increased demand for rural labor that raised rural wages and, together with declining food prices, reduced rural poverty.

Sustained, although much slower, agricultural growth in the 1990s reduced rural poverty to 26.3 percent by 1999/00. Since then, however, the slowdown in agricultural growth has become a major cause for concern. India's rice yields are one-third of China's and about half of those in Vietnam and Indonesia. With the exception of sugarcane, potato and tea, the same is true for most other agricultural commodities.

The Government of India places high priority on reducing poverty by raising agricultural productivity. However, bold action from policymakers will be required to shift away from the existing subsidy-based regime that is no longer sustainable, to build a solid foundation for a highly productive, internationally competitive, and diversified agricultural sector.

Issues and Challenges

It is here the challenge arises considering the implementation of the technology at various levels in the Global community. The need of the hour is not application of the technology but the adoption of appropriate technology, which would suit the particular level of the global community. In India, the farming practices are too haphazard and non-scientific and hence need some forethought before implementing any new technology.

Applications of agricultural inputs at uniform rates across the field without due regard to in-field variations in soil fertility and crop conditions does not yield desirable results in terms of crop yield. The management of in-field variability in soil fertility and crop conditions for improving the crop production and minimizing the environmental impact is the crux of precision farming.

Thus, the information on spatial variability in soil fertility status and crop conditions is a pre-requisite for adoption of precision farming. Space technology including global positioning system (GPS) and GIS holds good promise in deriving information on soil attributes and crop yield, and allows monitoring seasonally- variable soil and crop characteristics, namely soil moisture, crop-phenology, growth, evapotranspiration, nutrient deficiency, crop disease, and weed and insect infestation, which, in turn, help in optimizing inputs and maximizing crop yield and income. Though widely adopted in developed countries, the adoption of precision farming in India is yet to take a firm ground primarily due to its unique pattern of land holdings, poor infrastructure, lack of farmers' inclination to take risk, socio-economic and demographic conditions.

Factors Contribution to Decline of Agriculture

Slow Down in Agricultural and Rural Non-Farm Growth: Both the poorest as well as the more prosperous 'Green Revolution' states of Punjab, Haryana, Andhra Pradesh and Uttar Pradesh have recently witnessed a slow-down in agricultural growth and it ultimately lead for farmer's suicide. Some of the factors hampering the revival of growth are:

- Poor composition of public expenditures: Public spending on agricultural subsidies is crowding out productivity-enhancing investments such as agricultural research and extension, as well as investments in rural infrastructure, and the health and education of the rural people. In 1999/2000, agricultural subsidies amounted to 3 percent of GDP and were over 7 times the public investments in the sector.
- Over-regulation of domestic agricultural trade: While economic and trade reforms in the 1990s helped to improve the incentive framework, over-regulation of domestic trade has increased costs, price risks and uncertainty, undermining the sector's competitiveness.
- Government interventions in labor, land, and credit markets: More rapid growth of the rural non-farm sector is constrained by government interventions in factor markets — labor, land, and credit — and in output markets, such as the small-scale reservation of enterprises.
- Inadequate infrastructure and services in rural areas. Infrastructure is also a significant factor in the process of development but country like our rural Bharat has not posses the infrastructure such as roads, electricity, fertilizer and pesticides availability which caused the vulnerable damage to the growth of agriculture.

Weak Framework for Sustainable Water Management and Irrigation:

Inequitable allocation of water: Many states lack the incentives,Ø policy, regulatory, and institutional framework for the efficient, sustainable, and equitable allocation of water.

- Deteriorating irrigation infrastructure: Public spending inØ irrigation is spread over many uncompleted projects. In addition, existing infrastructure has rapidly deteriorated as operations and maintenance is given lower priority.
- Inadequate Access to Land and Finance: Stringent land regulations discourage rural investments: While landØ distribution has become less skewed, land policy and regulations to increase security of tenure (including restrictions or bans on renting land or converting it to other uses) have had the unintended effect of reducing access by the landless and discouraging rural investments.
- Computerization of land records has brought to light institutional weaknesses: State government initiatives to computerize land records have reduced transaction costs and increased transparency, but also brought to light institutional weaknesses.
- Rural poor have little access to credit: While India has a wide network of rural finance institutions, many of the rural poor remain excluded, due to inefficiencies in the formal finance institutions, the weak regulatory framework, high transaction costs, and risks associated with lending to agriculture.
- Weak Natural Resources Management: One quarter of India's population depends on forests for at least part of their livelihoods.
- A purely conservation approach to forests is ineffective: Experience in India shows that a purely conservation approach to natural resources management does not work effectively and does little to reduce poverty.
- Weak resource rights for forest communities: The forest sector is also faced with weak resource rights and economic incentives for communities, an inefficient legal

framework and participatory management, and poor access to markets.

- Weak delivery of basic services in rural areas: Low bureaucratic accountability and inefficient use of public funds: Despite large expenditures in rural development, a highly centralized bureaucracy with low accountability and inefficient use of public funds limit their impact on poverty. In 1992, India amended its Constitution to create three tiers of democratically elected rural local governments bringing governance down to the villages. However, the transfer of authority, funds, and functionaries to these local bodies is progressing slowly, in part due to political vested interests. The poor are not empowered to contribute to shaping public programs or to hold local governments accountable.

Measures Needed Areas

1. Enhancing agricultural productivity, competitiveness, and rural growth

   Enhancing productivity: Creating a more productive, internationally competitive and diversified agricultural sector would require a shift in public expenditures away from subsidies towards productivity enhancing investments. Second it will require removing the restrictions on domestic private trade to improve the investment climate and meet expanding market opportunities. Third, the agricultural research and extension systems need to be strengthened to improve access to productivity enhancing technologies. The diverse conditions across India suggest the importance of regionally differentiated strategies, with a strong focus on the lagging states.

Improving Water Resource and Irrigation/Drainage Management: Increase in multi-sectoral competition for water highlights the need to formulate water policies and unbundle water resources management from irrigation service delivery. Other key priorities include: (i) modernizing Irrigation and Drainage Departments to integrate the participation of farmers and other agencies in irrigation management; (ii) improving cost recovery; (iii) rationalizing public expenditures, with priority to completing schemes with the highest returns; and (iv) allocating sufficient resources for operations and maintenance for the sustainability of investments.

Strengthening rural non-farm sector growth: Rising incomes are fueling demand for higher-value fresh and processed agricultural products in domestic markets and globally, which open new opportunities for agricultural diversification to higher value products (e.g. horticulture, livestock), agro-processing and related services. The government needs to shift its role from direct intervention and overregulation to creating the enabling environment for private sector participation and competition for agribusiness and more broadly, the rural non-farm sector growth. Improving the rural investment climate includes removing trade controls, rationalizing labor regulations and the tax regime (i.e. adoption of the value added tax system), and improving access to credit and key infrastructure (e.g. roads, electricity, ports, markets).

2. Improving access to assets and sustainable natural resource use: Balancing poverty reduction and conservation priorities: Finding win-win combinations for conservation and poverty reduction will be critical to sustainable natural resource management. This will involve addressing legal, policy and institutional constraints to devolving resource rights, and transferring responsibilities to local communities.

Improving access to land: States can build on the growing consensus to reform land policy, particularly land tenancy policy and land administration system. States that do not have tenancy restrictions can provide useful lessons in this regard. Over the longer term, a more holistic approach to land administration policies, regulations and institutions is necessary to ensure tenure security, reduce costs, and ensure fairness and sustainability of the system.

Improving access to rural finance: It would require improving the performance of regional rural banks and rural credit cooperatives by enhancing regulatory oversight, removing government control and ownership, and strengthening the legal framework for loan recovery and the use of land as collateral. It would also involve creating an enabling environment for the development of micro-finance institutions in rural areas.

3. Strengthening institutions for the poor and promoting rural livelihood

Promoting Community-Based Rural Development: State Government efforts in scaling up livelihood and community-driven development approaches will be critical to build social capital in the poorest areas as well as to expand savings mobilization, promote productive investments, income generating opportunities and sustainable natural resource management. Direct support to self-help groups, village committees, user's associations, savings and loans groups and others can provide the initial 'push' to move organizations to higher level and access to new economic opportunities. Moreover, social mobilization and particularly the empowerment of women's groups, through increased capacity for collective action will provide communities with greater "voice" and bargaining power in dealing with the private sector, markets and financial services.

Strengthening Accountability for Service Delivery: As decentralization efforts are pursued and local governments are given more prominence in the basic service delivery, the establishment of accountability mechanisms becomes critical. Local governments' capacity to identify local priorities through participatory budgeting and planning needs to be strengthened. This, in turn, would improve the rural investment climate, facilitating the involvement of the private sector, creating employment opportunities and linkages between farm and non-form sectors

7

TRENDS IN RECENT AGRICULTURAL DEVELOPMENTS

INTRODUCTION

Agriculture provides basic sustenance to all living beings. It is very important that ecologically, socially and economically sustainable agriculture should become the backbone of the development process of the State. Agriculture should be sustainable so that the natural resources such as soil, water and biodiversity are used efficiently and equitably. It should be economically viable and lead to increasing employment opportunity, socially feasible, strengthening the role of women and other marginalized sections of the people. Equity in sharing benefits is vital for community participation in the conservation and enhancement of natural resources.

The Government is aiming to achieve 100% food security in the State and also to create avenue for export of agricultural produce for economic upliftment of the farming community. During the Tenth Plan period, the state is aiming an annual growth rate of 4% in Agriculture and 8% in Horticulture crops for sustainable Agricultural development, employment generation and poverty alleviation. The Government is focusing its policies towards overall development of agriculture sector in terms of increasing the cropping intensity by bringing every piece of land under cultivation, productivity increase, maximizing natural resources with parallel efforts to conserve them. The state aims at bring second green revolution in dry land Agriculture, while sustaining the tempo of agricultural development in irrigated agriculture. The change in land use pattern created serious concern among the agricultural planners to evolve suitable development strategies.

The agricultural performance of the state is given in the Table below. (Kg per ha.)

Crop	Average Yield		Highest yield in India	Tamil Nadu's place
	India	Tamil Nadu		
Rice	1804	3350	3510 (Punjab)	Second
Jowar (Cholam)	769	962	962 (Tamil Nadu)	First
Bajra (Cumbu)	610	1348	1348 (Tamil Nadu)	First
Redgram	616	710	1301 (Bihar)	Fourth
Total Food-grains	1562	2238	3830 (Punjab)	Fourth
Total Oilseeds	710	1611	1611 (Tamil Nadu)	First
Groundnut	733	1784	1784 (Tamil Nadu)	First
Cotton	193	305	410 (Punjab)	Third
Sugarcane	64562	106778	106778 (Tamil Nadu)	First

Source: Agricultural Statistics at a glance 2004 - Agrl. Statistics Division - GOI

The increasing trend of fallow lands (both current and other fallows) which was 22.93 L.Ha. during 1990-91 to 24.35 L.Ha. during 2001-02 which was further increased to 29.93 L.Ha. and 28.17 L.a. during 2002-03 and 2003-04 respectively due to drought situation caused reduction in cropping intensity from the average level of 120% to 113.4% during last 2 years. The gross cropped area declined to 53.16 L.Ha. during 2003-04, from the average normal coverage of 63 L.ha. leaving about 9.8 L.Ha. under fallow. Though the reduction of cropped area during 2003-04 is evidently more due to acute drought prevailed during that year, still there is reduction under gross cropped area to the tune of around 4 L.Ha. between 1990-91 to 2001-02. Inspite of various constraints such as fragmentation of land holdings, more than 52% of the cultivable area are under dryland condition, over exploitation of irrigation sources, deterioration of soil health due to continuous intensive cropping, shrinking trend of gross cropped area and also continuous severe drought prevailed in the past 2 years, Tamilnadu State continues to occupy top place in All India ranking towards the productivity of major crops.

Present Scenario

The present agricultural scenario are,

- Pattern of agricultural growth not uniform between region, crops and categories of farmers.
- New Technologies not permeated into dry land regions and unirrigated crops as expected.
- Number of dryland crops remain untapped to their potential
- Stagnation in growth of agricultural sector.
- Reduction of gross cropped area.
- Permanent and current fallows increased from 17.65 L.Ha. in 50's to 24.35 L.Ha. in 2001-2002.
- Depletion of water resources
- Degradation of land and soil quality
- Slow down in growth rate of major crops and yield of dry land crops.
- Yield gap exists between potential yield and actual yield of major crops.

SEASONAL CONDITION AND CROP PROSPECTS

The health, growth and normal yields of crops depend upon certain optimum conditions of rainfall, temperature, wind, cloudiness, etc., which normally occurs in a season. Magnitude of vulnerability of agriculture within a season again depend heavily upon weather variability within a season.

The Agricultural situation in Tamil Nadu largely depends on the quantum of rains received during seasonal rainfall of South West and North-East monsoon. Comfortable storage in Mettur reservoir coupled with uniform distribution of monsoon rains will bring good crop in delta region. But the State had faced consecutive drought situation for the past 2 years because of low and uneven precipitation.

During 2004-05, the state had experienced a dual situation, whilst it faced severe drought upto June-July 2004, it had torrential rains in the month of October leading to flood damage and submergence in some of the districts. The water from Mettur reservoir was released on 12.8.2004 after a delay of 2 months for Samba cultivation.

The area coverage under major crops showed a dip in coverage in the early part of the year, which however revived in the month of September 2004 onwards.

On 26th December, 2004 Tsunami waves which is unusual in the India Ocean attacked Tamil Nadu Coastal line and made fertile lands submerged with sea water. The crops raised in 4,796.10 Ha. were affected and the extent of area to be reclaimed is assessed as 3,664.24 Ha. and the total extent damaged in agriculture is 8,460.34 Ha. For relief to the crop damage, the Government have already ordered for the financial compensation of Rs.1.80 crores and a sum of Rs.16.96 crores towards the relief work to be taken up for reclamation of Tsunami affected agricultural lands. The Mobile Soil Testing Laboratories were deputed to Nagapattinam, Cuddalore and Kanyakumari districts to make preliminary assessment of soil damage. The damage caused by Tsunami are so complex and site specific and hence 100% soil survey of the damaged land in Nagapattinam was ordered and carried out. The Experts recommended soil amelioration by scrapping and removal of silty clay deposits, leveling, leaching with good water 3 or 4 times with short intervals to restore soil health, application of gypsum at 2 tonnes per hectare, liberal addition of organic manure or green manure. Mulching, raising salt tolerant crop/ varieties with 20% extra seed rate to ensure better germination.

The details of rainfall received during 2004 is as follows. (Unit: in mm)

Sl. No	Season	Normal Rain-fall	Actual Rainfall 2003	Actual Rainfall 2004	Deviation (%) w.r.to Normal	Deviation (%) w.r.to Last Year
1	Winter (January-February)	37.5	8.7	12.1	(-) 67.7	(+) 39.1
2	Summer (March-May)	127.5	124.9	283.4	(+) 122.3	(+) 126.9
3	Southwest Monsoon (June-September)	324.3	329.7	459.8	(+) 41.8	(+) 39.5
4	Northeast monsoon (October-December)	464.7	404.5	470.9	(+) 1.3	(+) 39.5
	Total	954.0	867.8	1226.2	(+) 28.5	(+) 41.3

Rabi Campaigns

In the era of globalization, Indian agriculture will face increasing competition. During initial period of Green Revolution agricultural extension played a crucial role in carrying the messages to the various parts of the State. However, with the increasing complexities in market and society, the extension methodology has to be sensitized to the new dimensions in agriculture. The task of extension is becoming more challenging in the context of globalization, which demands a system of market-linked extension. New approaches like precision farming knowledge management with modern information and communication technology are tools, which have to play crucial roles in strengthening sustainable agriculture of Tamilnadu and prepare it for facing competitive market.

Continuous drought for the past 2 years and also during early part of Kharif 2004 for the third year badly affected the agricultural situation in Tamilnadu. The gross cropped area declined to 53.16 L.Ha. from the average normal coverage of 63 L.Ha. leaving about 9.8 L.Ha. under fallow. The ground water level gone down to critical level and almost all the tanks and reservoirs went dry and posed threat to Agriculturists.

The receipt of adequate wide spread rains in late part of Kharif 2004 have created a conducive favourable situation for crop cultivation. Taking advantage of this situation, the department had taken sincere efforts to maximize the area coverage by creating enthusiasm among the farmers to take up cultivation through intensive campaign in all villages, at different stages.

- Immediately on release of Cauvery water in August, 2004, District Facilitation Centres were opened in the Delta districts and kept opened from 8.00 AM to 8.00 PM from 15.9.2004 and again 9.11.2004 to render all required assistance. Inputs were made available on war footing manner to bring maximum area under Samba. Intensive Preseason Campaigns were conducted in all the villages of delta districts.
- Phase I Intensive Rabi Campaign conducted in all the districts covering all the 17244 villages and 21029 campaigns conducted from 16th September 2004 to 15th October 2004. 4,55,336 farmers participated and benefited by these campaigns.
- Phase II campaign conducted at FIRKA LEVEL from 16th November 2004 to 31st December 2004 involving all line Departments such as Horticulture, Agricultural Engineering, Animal Husbandry, etc. besides Credit Institutes, leading input supply agencies involving people's representatives with mass media support. Inputs were made available at farmers doorstep to the

extent possible. In 1086 firkas, 1750 Nos. of Phase II campaigns conducted benefiting to 92209 farmers.

- Phase III campaign conducted from 16th January,2005 to 28th February, 2005.

The Preseason campaigns conducted at different phases based on the seasonality has really had very good impact among the farmers and welcomed by all. The Department also ensured supply of quality inputs to the farmers in time. These campaigns pave the way to the farmers to adopt new technology to get good yield.

The area coverage and estimated production upto 21.3.2005 under important crops are indicated below:

Sl. No.	Crops	Area (L.Ha.)		Production (L.MT.)	
		Target	Actual	Target	Actual
1	Rice	20.00	19.28	75.00	69.02
2	Millets	11.00	12.12	14.85	15.20
3	Pulses	7.50	8.46	3.38	3.66
	Total food-grains	38.50	39.86	93.23	87.88
4	Cotton (L.bales)	2.50	1.63	5.62	2.81
5	Sugarcane (Gur)	3.00	2.37	35.00	25.29
6	Oilseeds	10.00	8.64	17.00	13.91

Anticipating a favourable situation, the target proposed for 2005-06 is given in the Table below.

Sl. No.	Crops	Area (L.Ha.)	Production (L.MT)
1	Rice	20.00	78.00
2	Millets	11.00	15.51
3	Pulses	8.00	3.76
	Total Food-grains	39.00	97.27
4	Cotton (L.bales)	2.50	5.85
5	Sugarcane (Gur)	3.00	36.40
6	Oilseeds	10.50	18.59

Strategies

The strategies of the Government to achieve the goal in Agricultural sector are;

- Adoption of alternative cropping strategies derived for agro-climatic zone based cropping pattern evolved by TNAU to raise remunerative crops and to ensure maximum utilization of available land and water aiming to increase cropping intensity.
- As a part of Alternative Cropping Strategy, cultivation of Jatropha, Sugar beet and Sweet sorghum are aiming at, as announced by the Hon'ble Chief Minister on Contract farming basis through approved Industrial entrepreneurs for the production of Ethanol and Bio diesel. These crops are highly remunerative and can be grown in moderately fertile lands with comparatively less water.
- Taking up cultivation in vast tract of waste lands through Massive Comprehensive Waste Land Development programme as announced by the Hon'ble Chief Minister.
- Much emphasis on dry land development integrating Watershed Development and Waste land Development programme
- Adoption of dry land development technologies and crop production technologies for dry land crops to step up the productivity.
- Identification and promotion of relevant technologies to bridge the gap between the potential yield and actual yield of major crops.
- Speedy transfer of technologies to the farming community through ICT (Information and Communication Technology).
- Conduct of Seminars, Workshops and Intensive Pre-season campaigns involving line Departments to enthuse and assist farmers to harvest good crops and to get good profit.
- Efforts to provide technologies and advice on Agriculture, Horticulture soil and moisture conservation, Animal Husbandry etc., under one roof.

- Introduction of Contract farming system for Maize, Oilseeds, Pulses and Cotton in potential districts with assured buy back arrangements at pre-announced price or prevailing market price if it is high. The main objective of the programme is to ensure expansion of area under these crops and to motive farmers especially SF / MF farmers to get assured remunerative price for the produce.
- Ensure availability of quality seeds on enhanced SRR through Public Private Partnership.
- Massive adoption of integrated nutrient management and integrated pest management technology with emphasis on ecofriendly Agriculture Development.
- Much focus on restoration of soil through organic farming approach and to promote vermi composting, compost making through plueorotus, green manuring etc.
- Efforts to provide site specific macro and micro nutrient recommendations on the basis of soil test villagewise fertility Index.
- Promotion of micro irrigation to maximize water use efficiency.
- Empowerment of women by revitalizing TANWA groups.

STATE SCHEMES

Input Programmes

Seeds

Seed is a vital input that plays a critical role in increasing agricultural productivity. The productivity of crop largely depend on the genetic and physical purity of seeds. The Department of Agriculture has the primary responsibility of producing and distributing good quality certified seeds as per approved seed replacement rate. The Department of Agriculture has taken special initiative to address the following key areas to revitalize the quality seed production and supply.

- To achieve enhanced SRR through Public Private partnership
- To achieve National standards of S.M.R

- Proper utilization of valuable Breeder / Foundation seeds to enhance 'C' seed production.
- Promote seed village concept
- Focus to increase availability of quality seeds of dry land crops for area expansion and productivity increase.
- Strict enforcement of quality parameters for seeds
- Steps to enhance the efficiency of delivery mechanism
- Identification and introduction of High yield potential varieties released by other states / ICAR which are notified and suitable for Tamil Nadu conditions.
- Ensure advance planning on demand and supply by creating linkage with Department, Quasi Govt. Private seed producers.
- Strengthening of infrastructure for quality seed production and storage.
- Ensure all information on the portal such as availability, sale price, seed production / storage technologies, Data base on seed producer, seed dealers.

The following SRR are proposed to be adopted during 2005-2006, for Departmental Seed Multiplication Programme.

Crop	Seed Replacement Ratio
Paddy	17%
Millet	7%
Pulses	12.5%
Cotton	Rainfed 10% Irrigated 15%
Oilseedsa) Groundnut	5%
b) Gingelly	15%
c) Sunflower	50%
d) Castor	30%
e) Soya	20%

The main objective of the Seed Multiplication schemes undertaken by the Department are to produce and distribute Quality seed of Paddy, Millet, Pulses, Oilseeds & Cotton in time at reasonable price. The trading activity by the Department also helps to curb unnecessary price hike by the private seed traders.

The seed production activities are carried out in 3 stages.

i. Breeder Seed production by TNAU for State varieties and ICAR Centre for national varieties.

ii. The foundation seed production in 37 State Seed Farms, 5 State Oil Seed Farms and one pulses farm at Vamban.

iii. Certified seed production in farmers' holdings.The Department of Seed Certification is involved to ensure quality of seeds.

The Foundation and certified Seeds produced in Government Seed Farms and farmers' holdings respectively are processed in 16 Major 2 Medium and 56 Mini Seed Processing Units in the State with annual capacity of 28500 MT. The entire seeds produced are distributed to the farmers through 880 Agricultural Extension Centres located in the State. The Quantity Seeds of Paddy, Millet, Pulses, Cotton and Oilseeds procured and distributed upto February 2005 and the target proposed for 2005-06 is as follows:-

(In MT)

Crop	2004-2005		Programme for 2005-2006
	Target	Achievement	
A. Procurement			
1. Paddy	17000	11341	17000
2. Millets	400	340	400
3. Pulses	1900	1064	1900
4. Oilseeds	8442.3	2562	8708
5. Cotton	367	125	367
B. Distribution			
1. Paddy	17000	10465	17000
2. Millets	400	340	400
3. Pulses	1900	1181	1900
4. Oilseeds	8442.3	2933	8708
5. Cotton	367	137	367

The Government have made substantial allocation for implementing seed multiplication schemes and the details are as follows:-

(Rs. in lakhs)

(Rs. in lakhs)			
Crop	BE 2004-05	RE 2004-05	Achmt. upto February 2005
Paddy & Millet	1953.44	2063.460	1601.691
Pulses	420.06	401.40	250.586
Oilseeds	602.34	601.140	515.250
Cotton	119.71	93.650	61.480

During 2004-2005, as announced in the Assembly, 100 Agricultural Extension Centres were repaired at a cost of Rs.100 lakhs for proper safe stocking of seeds to ensure distribution of quality seeds to the farmers.

For drying and also for protecting seeds, 213 Nos. of tarpaulins were purchased at a cost of Rs.10 lakhs.

Fertilizer

The Department of Agriculture shoulders the responsibility to closely monitor the production and supply of fertilisers to ensure timely availability to the farmers. After continuous drought situations for the past more than 2 years, during 2004-2005, extensive cultivation was resorted to by the farmers due to favourable seasonal rains from September 2004. This has created greater demand for fertilizers especially urea. This sudden spurt in demand was effectively tackled by securing additional quantity of 2.38 L.MT of urea during Rabi till November 2004 despite extreme limitations and timely availability ensured. A public notice was released in the dailies sternly warning the traders against indulging in over pricing or anyother malpractice and also inviting complaints from the farmers on non availability, sale at higher price. To enforce this warning 608 fertilisers shops were inspected and 79 stop sale orders issued 42 nos.of warning issued and 59 nos.of license suspended temporarily. To monitor the fertilizer supply, Facilitation centres were opened in all the districts from 8.00 A.M to 8.00 P.M. and the Department ensured timely availability of fertilizer at correct price.

The progress on supply of chemical fertilisers during 2004-05 and programme for 2005-06 are as follows:-

(Unit in LMTs)

Nutrient	Target 2004-05	Achievement upto February 2005	Programme for 2005-06
Nitrogen	5.40	5.18	5.40
Phosphorus	2.31	2.26	2.31
Potash	2.24	2.38	2.24
Total	9.95	9.82	9.95

Micronutrients

Micronutrients are very important for maintaining the Soil Health and also for increasing productivity of crops. Micro Nutrient deficiency is wide spread as evidenced in the increase in the percentage of Soil exhibiting deficiency of micro nutrient over years. Poor adoption of Micro nutrient Management is another reason for stagnation of Agricultural Productivity. The adoption of Intensive cultivation practices causes depletion of Micronutrient levels in most of the soil. The Department of Agriculture is constantly advocating balanced application of Micronutrients to the soil suitable for different crops. 18 Atomic Absorption Spectro Photo meter (sophisticated equipment to test micro nutrient content in the soil) have already been provided in the Soil Testing Laboratories. To meet the soil testing requirement of Sivagangai and Virudhunagar, during 2004-2005, 2 Nos. of Atomic Absorption Spectro Photometer were purchased for the above 2 districts also and by this, all the districts have been extended with facilities for testing the soil for Micro nutrient content. To make available quality Micro Nutrient mixture through the Departmental outlets, one Micro Nutrient Production Unit is functioning at Kudumianmalai to produce and supply 1250 MT of Micronutrient mixtures annually for Paddy, Millets, Pulses and Oilseed crops.

During 2004-05, upto February 2005, 1192 M.T of Micronutrient mixture have been produced and 1094 MTs were distributed. This programme will be continued during 2005-06 also.

Fertilizer Control Laboratories

The Govt. is very keen to make available quality fertilizers to the farmers as per the standards prescribed under the Fertilizer Control Order, 1985.

Strict enforcement work are being done by drawing samples from wholesale and retail sale outlets as well as from the mixing units. There are 14 notified Fertilizers Control Laboratories functioning in the State, with the analyzing capacity of 16,800 samples per annum. During 2004-2005, upto February 2005, 15,603 Nos. samples were analyzed. During 2005-2006, 16800 Nos. of fertilizer samples will be analysed by the laboratories to ensure quality.

Organic Farming

Due to continuous cropping, indiscriminate use of chemical fertilizers and inadequate application of organic manure to the soil, the general soil health of the State is getting deteriorated. The organic manure content in the soil has gone down from 1.20% in 1971 to 0.68% in 2002 in Tamilnadu, because of less use of organic inputs.

The decline in organic matter content has made undesirable changes in soil biodiversity and disruption in harmony of crop plants, which affects soil fertility and productivity. Considering its importance, the application of Bio-fertilizer, Green Manure, Green Leaf Manure, Vermi Composting, Composting of farm wastes through Pleurotus is popularized as a part of Integrated Plant Nutrient Management Technology. To ensure the adoption of the organic farming technology the following programmes are implemented.

Composting of Farm Waste through Pleurotus

During 2004-05, 2000 Nos. of kits each containing 1 Kg of Pleurotus, 5 Kgs of Urea and one Pamphlet have been distributed to the farmers at free of cost with an expenditure of Rs.2.80 lakhs to make compost by utilizing farm waste through Pleurotus for field applications. This programme for the distribution of 2000 No. of kits at a cost of Rs.2.80 lakhs will be continued during the year 2005-06 also.

Scheme to Promote vermi Composting

Under Vermi Compost Scheme, 100 Nos. of demonstration, cum trainings benefiting totally 5000 farmers at a total cost of Rs.3.85 lakhs has been conducted during 2004-05. The Scheme will be continued during 2005-06 also. The Department is also taking efforts to produce vermi compost in the Govt. farms for distribution to the farmers to the extend possible through AECs.

Biofertilizer Production and Distribution

To increase the availability of Bio-fertilizers, the Production of Bio-fertilizer inoculum packets containing Rhizobium, Azospirillum and Phosphobacterium are taken up in 6 Biofertilizer Production Units functioning at Cuddalore, Ramanathapuram, Salem, Kudumianmalai, Sakkottai and Trichirappalli with the Annual Production capacity of 1400 Mts.(70 L.Nos. of pockets of 200 gms each). The quality of biofertilizers produced in these units is monitored through Biofertilizer quality control Laboratory functioning at Trichirappalli.

The Biofertilizer Production Programme for 2005-06 is as follows:

Sl.No	Name	Quantity (MT)
1	Rhizobium	280
2	Azospirillum	800
3	Phosphobacterium	320
	Total	1400

During 2004-2005 upto February 2005, a quantity of 1109 MTs biofertiliser were produced and 940 M.T. were distributed to the farmers through Agricultural Extension Centres. This programme will be continued to produce and supply 1400 M.T during 2005-06 also.

Blue Green Algae and Azola

The Blue Green Algae and Azolla are reported to act as a supplementary sources of Nitrogen and have the capacity to fix atmospheric Nitrogen and made available to the crop. To make the inoculum available to the farmers, 500 M.T of Blue Green Algae and 500 M.T of Azolla are produced in the Govt. farms and supplied to the farmers.

Scheme to Promote Green Manuring

Raising of Green manure crops and ploughing insitu will enhance organic content of the soil. This practice has suffered by the intensive cultivation of other field crops.

As there is considerable decrease of animal husbandry in the farms, the availability of Farm yard Manure is felt inadequate and hence use of green manuring and green leaf manuring are the alternative source that could add organic matter to the soil.

Hence to motivate and to encourage the farmers to raise the Green manure crop, a state scheme is under implementation at a cost of Rs.25.00 lakhs to produce 125 M.T of Green Manure seeds and to distribute at 25% subsidy. During 2004-05, 137 M.T of green manure seeds have been produced upto February 2005 and 117 M.T Green Manure Seeds have been distributed at 25% subsidized cost.

This programme will be continued during 2005-06 also.

Popularising INM (Integrated Nutrient Management) technologies among farmers through Farmers Training Centres

Resource exhaustive crop varieties popularized during the Green revolution phase have brought in question of soil nutrient sustainability. While major nutrients are applied to soil through inorganic fertilizers, continuous mono-cropping takes up cumulatively higher quantities of other minor nutrients that are not replenished. In the absence of organic manure application, the soil profile is altered. Application of large doses of chemical fertilizers in an imbalanced way not only polluted environment and soil but also not given economic returns in terms of yield.

Hence to keep soil health and productivity adopting Integrated Nutrient Management technology has become highly essential. This technology is being popularized through training programme to farmers in the 12 farmers training centres.

This training programme was introduced to train 2700 farmers during 2003-2004 at a cost of Rs.8.64 lakhs. This training programme have been enlarged during 2004-2005 and totally 13500 farmers were trained at the rate of 500 farmers per district at a cost of Rs.43.20 lakhs. Each training was conducted for 2 days and 50 farmers were trained per batch. Totally 270 trainings were conducted.

The achievement is as follows:

Sl. No.	Details	Target	Achmt. upto February 2005
1.	INM - No.of training conducted (Nos.)	270	270
2	No.of farmers trained	13500	13500
3.	Financial (Lakhs)	43.20	43.20

SOIL MANAGEMENT AND SOIL TESTING

Soil Survey and Land Use Organisation

The objective of the scheme is conducting soil survey and preparation of soil resources inventory including the nature of soil occurring in the area, their morphological, physical and chemical characteristics through field studies and laboratory analyses, classifications according to internationally recognized system of soil classification and mapping their extent on standard topographic base maps and finally interpreting for variety of uses. The above survey is being taken up through the four Soil Survey units at Coimbatore, Thanjavur, Vellore and Tirunelveli. During 2004-05 an area of 0.99 L.Ha. has been surveyed as against the target of 1.10 lakh ha. till February 2005. An area of 1.10 L.Ha. will be covered by "detailed soil survey" by these units during 2005-2006.

Reclamation of Saline and Alkaline Lands

The Saline and Alkaline Land Reclamation Scheme is functioning to reclaim problem soils in Nagapattinam, Thiruvarur, Kancheepuram, Thiruvallur, Vellore, Thiruvannamalai, Salem and Namakkal districts in an area of 4000 ac. During the year 2004-2005, an area of 2886 acres were reclaimed up to February, 2005. It is programmed to reclaim 4000 acres during 2005-2006 under this scheme.

Soil Testing Laboratories

As a part of Integrated Nutrient Management technology, the farmers have to be motivated to apply need based Macro / Micro nutrients to the crops, to ensure economy in input use for higher productivity and profitable return.

To advise the farmers on site specific fertilizer recommendations on the basis of native soil fertility, analysis of soil samples is being carried out through 19 Soil Testing Laboratories and 16 Mobile Soil Testing Laboratories functioning in the State.

A total number of 4.13 Lakh of soil samples have been analyzed through Soil Testing Laboratory against the target of 5.28 L.Nos. and a total number of 1.71 Lakh Nos.of soil samples through Mobile Soil Testing Laboratory were analysed till February 2005 against the targetof 2.82 L.Nos.

The above laboratories also suggest suitable reclamation measures for the problem soils, preparing village level fertility index and assessing the quality of irrigation water. The Mobile Soil Testing Laboratories are the effective propaganda aids to disseminate the soil, bio-fertilizers and micronutrient management technologies among the farming community through Village Adoption Programme, Soil Testing campaigns, Exhibitions, etc.

Moreover, there is an apex organization viz. "Central Control Laboratory" located at Kudumianmalai, which inculcates technical competence through training to the laboratory personnel and also monitors accuracy of analysis of all the laboratories in the Department of Agriculture.

PLANT PROTECTION

Pest Surveillance

The Integrated Pest Management Technologies are recommended and adopted extensively by the farmers. The success of the Integrated Pest Management largely depends on constant monitoring of pest and disease situation in the field, defender population, etc. To augment its efforts the pest surveillance are done in 7356 locations every week and forewarning messages are given to the farmers on the possible out break. Because of the continuous surveillance, the serious pests like paddy stemborer, leaf roller, Brown Plant Hopper, groundnut red hairy catter pillar, prodenia, cotton boll worm, etc. which caused severe damages a decade ago were kept under control.

Crop and Plant Protection Scheme

The Plant Protection measures were programmed to be taken up during 2004-05 to cover an area of 37.5 lakh ha. in food crops, 18.2 lakh ha in non food crops and seed treatment measures in 22.0 lakh ha. and upto February, 2005 an area of 31.760 Lakh ha. were covered in food crops, 15.132 Lakh ha. under non food crop and 22.693 L.Ha. under seed treatment. This will be continued during 2005-06 also.

Under this scheme, Government have sanctioned funds for the procurement of bio-control agents and bio-pesticides for distribution to the farmers at subsidized cost through Centrally sponsored Schemes such as Cereals Development Programme, SUBACS, ICDP Cotton, ISOPOM-Pulses, Oil Seeds and Maize.

Integrated Pest Management (IPM)

The Integrated Pest Management Technology to combat pest and diseases was introduced in Tamilnadu through FAO

Programme in late 80's and special thrust were laid to this concept through Farmers Field School approach. This season long training cum demonstration component involving 30 farmers in each training, created greater awareness among the farmers and being practiced in almost all the villages. Because of the large scale adoption of IPM technologies, the technical grade pesticide consumption which was annually around 10,000 MTs. a decade ago has come down to around 1300 MTs.

During 2004-05 the following programmes are implemented under IPM through Centrally Sponsored Schemes.

Sl. No.	Name of the component	Physical (Nos.)	
		Programme	Achievement upto Feb 05
1	IPM- Demonstration cum FFS under Cereals Development Programme	330	330
2	IPM Demo cum FFS under Cotton Mini-mission II	350	340
3	IPM Demo cum FFS under ISOPOM maize	16	16
4	IPM Demo (FFS) under ISOPOM - Oilseeds.	66	65
5	IPM Demonstration under ISOPOM Pulses.	25	24

This IPM approach will be continued during 2005-06 also.

Bio-control Agents Production Centre & State IPM Centre

As a part of IPM activities, production of Bio control Agents are resorted by the Department and distributed to the farmers.

There are 12 State Bio-Control Agents Production Centres and 51 Parasite Breeding Centres functioning in this State to take the production of the Bio Control Agents as indicated below:

Sl. No.	Bio control agent / Crop / Pest	No .of Centres	Area Coverage (In Ha.)
1	Trichogramma Chilonis (Sugarcane Internode borer)	21	10500
2	Bethylids, Branchonids and Eulophids (Coconut black headed caterpillar)	21	10250
3	NPV (Cotton Boll Worm)	3	1500
4	NPV (Groundnut red hairy Caterpillar and Prodenia)	4	480
5	Green Muscardine fungus tubes (Coconut Rhinocerous beetle.)	2	10000
6	Bio-control agents (Cotton, Paddy, Pulses, Groundnut)	12	60000

This scheme will be continued during 2005-06 also.

Pesticides Testing Laboratories

Nine Pesticides Testing Laboratories are functioning in the State with the capacity to analyse 13860 samples per annum.

These notified laboratories are functioning to enforce Insecticide Act and rules. During 2004-2005, 12721 Nos. of samples were analyzed up to February 2005. It is programmed to analyze 13860 Nos. of Pesticide samples during 2005-2006.

CROP ORIENTED STATE SCHEMES

Pulses Village Scheme

India is the largest producer and consumer of pulses with 33% of the World"s area and 22% of the production. In India, Tamilnadu contributes 3.2% of the total area and 2.42% of production of pulses.

There is big gap exists between demand and production of pulses. The average productivity of pulses for the past 5 years, is around 400 kgs./Ha.

The major reason for stagnation in productivity is that more than 90% of the pulses area is under rainfed condition and raise in marginal soils. Hence, a Pulses Village scheme is under implementation since 2003-04 with the objective to enhance the pulses productivity from 450 kgs./ha. to 750 kgs./ha.

This scheme is under implementation during 2004-05, in the 20 districts of Thiruvannamalai, Pudukkottai, Trichy, Perambalur, Karur, Thanjavur, Thiruvarur, Nagapattinam, Salem, Namakkal, Dharmapuri, Dindigul, Theni, Virudhunagar, Thirunelveli, Thoothukudi, Cuddalore, Erode, Vellore and Coimbatore where pulses are grown extensively to cover 280 villages at the rate of 14 villages in each district.

In each village 100 acres (40 Ha.) are being covered under Pulses totaling to 28000 Acres (11200 Ha.)

This will serve as a compact block demonstration for the neighbouring farmers to know and adopt the latest technologies.

In these demonstration plots, pulses seeds are supplied at 50% subsidy to the maximum of Rs.15/kg. besides the supply of DAP at 4 kgs. per Ac. with a maximum subsidy of Rs.20/Ac. supply of Dhal Processing Units to the maximum subsidy of Rs.10,000 each is also extended .

This scheme is under implementation at a total cost of Rs.41.20 lakhs.

The component wise physical and financial target during 2004-05 and achievements upto February, 2005 are as follows:

Sl. No.	Component	Unit	Physical		Financial (L.Rs.)	
			Target	Achmt. upto Feb.05	Target	Achmt. upto Feb.05
1.	Compact Block Demo	(Ha.)	11200	8560	--	...
2	Seed Distribution	(MT)	224	172	33.60	29.941
3	D.A.P. Spraying	(MT)	112	51.62	5.60	3.90
4	Dhall Processor	Nos.	20	10	2.00	1.00
	Total				41.20	34.84

COCONUT DEVELOPMENT PROGRAMME

In Tamil Nadu, Tall, TxD and DxT seedlings are produced in 22 coconut nurseries with two mother palm gardens to make available the required number of coconut seedlings to the farmers.

The seedlings are sold to the farmers at Rs.12/- per Tall seedling Rs.20/- per Tall x Dwarf seedling and Rs.75/- per Dwarf x Tall seedling.

During 2004-05 the production of Dwarf x Tall Hybrids is less due to limited availability to Mother Palms in Government Farms. The progress made under procurement of seed nuts and distribution of coconut seedlings during 2004-05 and programme for 2005-06 are given below: (Unit: L.Nos.)

Sl. No.	Variety	Target 04-05		Achievement 04-05		Programme 05-06	
		Nuts procurement	Seedlings Distribution	Nuts Procurement	Seedlings Distribution (Feb. 05)	Nuts Procurement	Seedlings Distribution
1.	Tall	3.0	2.25	4.271	2.445	3.00	2.25
2.	T x D	3.20	2.24	2.592	2.143	3.20	2.24
3.	D x T	1.00	0.35	0.092	0.016	0.200	0.070
4.	**	5.30	4.00	..	..	7.80	6.00

** Free Seedlings distribution to female infants

The Programme will be continued during the year 2005-06 also.

COTTON DEVELOPMENT SCHEME

The area under cotton which was around 2.67 L.Ha. during 1992-93 has declined to around one L.Ha. during 2003-04. The decline has been due to failure of monsoon and switch over by farmers to other less risky and more profitable crops like maize.

The mills require a large quantity of cotton at cheaper rate. At present, they incur heavy expenditure on transport from neighbouring states. The farmers therefore need to be motivated to increase the area under cotton and to adopt the latest technologies.

Much thrust is being laid to increase the area of cotton through sustained extension efforts backed with supply of quality inputs. There has been possibility of increasing area under irrigated and rice fallow cotton, during current year.

The rainfall received during June-December coincides with rainfed as well as winter irrigated seasons, which provides avenues for expansion of area under winter cotton. Steps are being taken to increase the area by motivating farmers to go for short duration varieties which is a remunerative alternative crop for rice fallow pulses. To encourage cotton growers in Tamilnadu contract farming is encouraged with the buy back arrangements for their produce which is detailed in a separate chapter. In order to increase area, production and productivity of cotton in Tamil Nadu, the Department of Agriculture is adopting the following strategies and schemes.

- Cultivation of hybrids
- Field demonstration on IPM and frontline demonstration on machineries and implements
- Micro Irrigation systems to maximise water use efficiency.
- Farmers Training on Production Technologies.
- Constant monitoring of pest and diseases.
- Promoting Inter cropping / mixed cropping/ Rice fallow cotton.
- Contract farming among the farmers with reputed mills.
- Insurance cover / crop loans.
- Involving Self Help Groups of farmers for area expansion.

The progress made under this scheme upto February, 2005 and the programme for 2005-06 are given below:

Sl. No.	Particulars	2004-05		Programme for 2005-06
		Target	Achievement upto Feb. 2005	
1.	Area covered under improved varieties/hybrids (L.Ha.)	2.50	1.630	2.50
2.	Lint production (L.Bales)	5.62	2.809	5.85
3.	Certified seed production (Qtl.)	3670	1250	3670
4.	Certified seed distribution (Qtl)	3670	1370	3670
5.	Seed procurement (financial: L.Rs.)	93.65	61.48	
6.	NPV Distribution (Ha.)	1500	1300	1500
7.	NPV Distribution (financial: L.Rs.)	0.60	0.532	0.60

This scheme will be continued during 2005-06 also.

STATE SCHEME ON PEST SURVEILLANCE IN COTTON

The scheme is implemented with the objective to monitor the incidence of pest and diseases, in the cotton growing areas in the 10 selected districts, through scouting thereby reducing crop losses by adopting IPM measures.

The village unemployed youth are trained in pest surveillance activities and they act as scouts for monitoring the pest incidence at village level. A scout has to cover 100 ha. and an assistance of Rs.500 is given for scout per month and totally Rs.3000/scout is given for the crop duration.

The scheme was announced by the Hon'ble Minister for Agriculture during 2002-03, and from 2002-03 onwards it is being implemented in the State. During 2004-05, an amount of Rs.1.50 lakhs has been sanctioned to cover 5000 ha. under the scheme.

Sl. No.	Component	Target	Achmt. upto February 2005	Programme for05-06
1.	Physical	5000 Ha.	3000	5000 Ha.
2.	Financial	1.50 L.Rs.	0.80	1.50 L.Rs.

CONTRACT FARMING IN COTTON

The cotton is cultivated in Tamil Nadu in an area of 2 lakh ha. in the normal years and the state production is around 4 - 5 lakh bales. More than 60% of the textile mills in the country are located in Tamil Nadu, which annually consumes nearly 50 lakh bales of cotton. The state production meets only 10% of the requirement. This huge gap between demand and supply has forced the mills to get their requirement from other states.

Since, there is a huge demand for cotton within Tamil Nadu and to link farmers with the assured marketing and to protect the interest of both the farmer and the industry, contract farming is promoted by the State Government through Department of Agriculture.

The farmer is provided support in diverse areas such as marketing, input, technical assistance, credit, production management, insurance coverage etc., under contract farming.

Contract Farming strategy has been introduced by private firms M/s. APPACHI Cotton Company of Pollachi and Super Spinning Mills Ltd. Coimbatore.

The firms have arranged input distribution, insurance benefits, crop loan, with buy-back guarantee to the farmers covered under the Programme. State Government has ordered waiver of 1% market cess for M/s APPACHI Cotton Company for three years from 2002-03 and proposal for waiver of market cess to the Super Spinning Mills is under consideration of the Government.

The progress made sofar is as follows:-

Appachi CARE Foundation, Pollachi				
Year	Area covered (Ac.)	Varieties cultivated	Farmers benefited (Nos.)	Districts
2002-03	916	Bunny, RCH 2, Supriya	423	Coimbatore, Theni, Namakkal.
2003-04	710	Bunny, Surabhi	368	Coimbatore, Namakkal, Ramnad
2004-05	753	Bunny, SVPR3, MCU 7, Spic 22, RCH 3600, MCU 7.	749	Tiruvarur, Thanjavur, Nagapattinam.
Total	2379		1540	

Super Spinning Mills, Coimbatore				
Year	Area covered (Ac.)	Varieties cultivated	Farmers benefited (Nos.)	Districts
2002-03	1281	Sara 1, Sara 2, MCU 5.	820	Coimbatore, Salem.
2003-04	2558	MCU 5, Sara 2, TCHB 213, Bunny.	1310	Salem, Coimbatore, Vellore.
2004-05	2208	Surabhi, MCU 5, MCU 7, Spic 22, Bunny.	566	Namakkal, Salem, Theni, Madurai, Tiruvarur, Nagai.
Total	6047		2696	
Grand Total	8426		4236	

The Programme for 2005-06 is as follows:-

Name of Mills	Proposed area coverage (acres)
Appachi Cotton company	2500
Super Spinning Mills	20000
Royal Classic Mills	20000

Tripartite Contract Farming

During 2004-05, the Department of Agriculture has organized a 'Cotton Stake Holders Meet to promote contract farming' at Coimbatore on 20.11.04 involving officials from GOI, mills, cotton farmers, CCI, Credit Institutions, Insurance Agencies, Marketing Sector, Inputs Suppliers, TNAU and

Central Institute for Cotton Research. The Commissionerate of Agriculture has formulated a tripartite contract farming model for enhancing cotton production in the state and approached various agencies viz., Mills, Cotton Corporation of India, NAFED, Insurance Agencies, Banking Institutions for participation in the Programme. The main objectives of the model developed by the Commissionerate of Agriculture are as follows:

- To focus on cluster of villages.
- De-risk the farmers
- Viable price mechanism
- Credit linked inputs supply
- Insurance Coverage
- Ensure quality as per mills specifications
- Extension support by the Department

Under this model, the Department of Agriculture shall act as the Chief Co-ordinator of the Programme and TNAU to render the R&D support. Dispute settlement, if any, shall be looked after by CICR.

Under this programme, Royal Classic Mills, Palani have consented to take up contract farming in cotton during summer cotton season 2005, in the districts of Salem, Namkkal, Dindigul and Erode in an area of 5000 acres projecting the production of 50,000 Qtls and agreed to procure cotton at Rs.1850/- per qtl. or on the prevailing market rates whichever is high. The Cotton Corporation of India will act as the alternate procurement agency if the Mill fail to procure the produce. Rasi Seeds (P) Ltd., Salem, consented to supply RCH 2Bt. Seeds to the farmers covered under contract farming Programme. Royal Classic Mills so far distributed 1300 nos. of seed packets to identified 1300 nos. of farmers and imparting training to field functionaries of the districts is under progress. The Royal Classic Mills propose to cover about 20,000 Acres in 6 districts during 2005-06. The Super Spinning Mills now came forward and willing to join the Tripartite model of contract farming by covering 4,900 acres in Madurai, Salem, Erode and Theni districts during Summer 2005. The Super Spinning Mills proposed to cover about 20,000 acres during 2005-06 in 10 districts.

Our State model has been accepted by the GOI with laurels. Not only praising the efforts taken by our State, the GOI has circulated this Tripartite model of Contract Farming to all the States for emulation and also suggested that such an arrangement may augment the tradition of contract farming of cotton so that the cotton of desired quality is available to mills and also indicated that the memorandum of agreement may be helpful to State extension machinery as well as to the organization aspiring to popularize the contract farming in cotton.The Commissionerate of Agriculture will provide extension support as well as training to the farmers. Awareness campaigns will be conducted at village level. Commercial Banks will provide credit facilities to the identified farmers with Insurance coverage.

Contract farming through NAFED for Maize, Pulses and Oilseeds

The Hon'ble Chief Minister has announced on 5.6.2003 for adoption of alternative cropping strategy and crop diversification for delta areas wherein raising less water intensive crops were focused. The Department is taking all possible efforts to bring substantial area under less water intensive crop. The success of the alternative cropping strategy will largely depend on involvement of more number of farmers to raise less water intensive crop. The Commissionerate of Agriculture has introduced contract farming approach for maize, oilseeds and pulses with the following objectives

1. Area expansion under less water intensive crops
2. Increase in cropping intensity
3. Farmers income augmentation
4. Reduce middle men role in marketing
5. Increase in production and productivity

Under this programme NAFED will purchase the produce through buy back arrangements at prevailing market price or pre announced price whichever is higher. The Department of Agriculture will motivate and identify the farmers for this programme besides providing technical knowledge to increase the production and supply of inputs to the extent possible.

The Contract Farming Programme is proposed to be implemented as follows: (Unit in MT)

Crop	District	Quantity proposed to be procured
Maize	Theni, Dindigul	3000
Groundnut	Cuddalore, Vellore, Tiruvannamalai, Villupuram, Salem	2000
Gingelly	Delta Districts	500
Pulses	Delta Districts of Thanjavur, Thiruvarur and Nagapattinam	5000

This programme is commenced from January, 2005 and will be implemented till June, 2005 in about 500 villages during irrigated season. About 30,000 farmers are expected to be involved under this programme. This profitable marketing arrangement is expected to attract more farmers which will achieve the dual benefit of ensuring area expansion under less water intensive crop and to fetch good price to the farmers.

Programme for Jatropha, Sweet Sorghum and Sugarbeet Cultivation

The Hon'ble Chief Minister of Tamil Nadu has announced a massive scheme for cultivation of Jatropha, Sweet Sorghum and sugar beet as a part of Alternative Cropping Strategy on contract farming basis for the production of ethanol and bio diesel. This scheme is proposed to be taken up linking Industrial entrepreneurs with buy back arrangements at pre-fixed price for the produce. These crops are less water intensive and highly remunerative to the farmers.

Promotion of Jatropha

The Promotion of Jatropha cultivation by contract farming is entrusted to 4 entrepreneurs identified by the Government by allotting specific district of jurisdiction to them as follows.

Firm	District allotted	Action Programme (Ha.) 2004-05	2005-06
M/s. Mohan Breweries and Distilleries Limited	Thiruvallur	5000	80000
M/s. Shiva Distilleries Limited	Karur, Dindigul, Erode and Coimbatore	2100	3800
M/s. Dharani Sugars and Chemicals Limited	Virudhunagar	Firms yet to make their commitment	
M/s. Riverway Agro Products Private Ltd.	Tirunelveli and Thoothukudi		

The extent of Jatropha cultivation organized by M/s Mohan Breweries and Distilleries Limited is 119 Ha. and M/s. Shiva Distilleries Limited is 180 Ha. upto February 2005, as against the programme to cover 5000 Ha. and 2100 Ha. respectively. Due to non-availability of quality seed material, the area expansion gets delayed. The firms have taken steps to obtain quality seeds from Chattishgarh and Rajasthan and Nationalised banks were being persuaded by them to get credit facility. The preparations of Bipartite and Tripartite agreements is in progress to enter into the Contract Farming. The procurement rate is committed as Rs.5000/- per tonne. M/s. Mohan Breweries Limited have proposed to establish an extraction cum transesterification unit. The Government have announced that there will be no purchase tax on Jatropha. Sales tax on bio diesel from Jatropha seeds will be 4% besides issue of special incentives as per industrial Policy statement 2003 of the Government of Tamil Nadu.

SCHEME ON ELITE PLANTING MATERIAL AND MODEL JATROPHA PLANTATION

The NOVOD Board have sanctioned Rs.150 lakhs (Rs.75 lakhs for Department of Agriculture and Rs.75 lakhs for TAWDEVA) for implementing the project on "Elite planting material and model Jatropha Plantation". Under this programme, the model Jatropha plantation will be established to cover 600 Ha. in the Government farms belongs to Agriculture and Horticulture Departments @ Rs.25,000/Ha. as 100% assistance towards inputs, equipments, planting materials, maintenance, etc. The Chief Engineer(Agricultural Engineering) will assist in landscaping, digging 25 bore wells and installation of Drip irrigation in these identified farms based on the need. 720 Kgs of Jatropha curcas seeds were obtained from Indira Gandhi Agricultural Uniersity Raipur and 900 Kgs. from Tamil Nadu Agricultural University, Coimbatore for raising seedlings. Besides, 25000 nos. of seedlings have been obtained from the Tamil Nadu Agricultural University, Coimbatore to take up planting in Annapannai. This programme launched on 25.2.2005 and planting commenced.

During 2005-2006 NOVOD Board will be approached to sanction of Rs.15 lakhs as maintenance subsidy for these plantations. These plantations will serve as modal plantation. The seed material to be obtained from these plantations will be utilized for further area expansion.

PROMOTION OF SWEET SORGHUM

The firms . M/S Mohan Breweries & Distelleries Limited, M/S Bannari Amman Sugars Ltd., M/s.Shree.Harichandana Estates Pvt, Ltd., M/s.Dharani Sugars & Chemicals Ltd., M/s.Shree. Ambika Sugars Ltd., have furnished to Government their expression of interest for entering into contract farming agreements with farmers in selected districts, who will cultivate Sweet Sorghum and sell Sweet Sorghum stalks to the Processing Unit of the firms concerned, and Government have permitted the firms to take up the programme in the following districts. In addition, Thiruarooran Sugars Limited have also furnished their programme.

Firm Name/Districts permitted by Govt.	Districts Programmed	Action Plan (acres)	Seed requirement (tonnes)	Area covered upto Feb 05 (acres)
M/s.Mohan Breweries &	Kanchipuram	100	0.300	7
Distilleries Ltd Kanchipuram	Villupuram	100	0.300	1
Villupuram	Tiruvallur	100	0.300	45
	Total	300	0.900	53
M/s.Dharani Sugars &	Virudhunagar	100	0.300	1.5
Chemicals Ltd.,Virudhunagar	Tirunelveli	800	2.400	1.0
	Total	900	2.700	2.5
M/s.Shree Ambika Sugars Ltd.,	Cuddalore	50	0.150	102.5
Cuddalore	Perambalur	20	0.060	-
Perambalur	...	....	....	...
Tiruvannamalai	...	...	...	...
Pudukkottai	Salem	10	0.030	-
Thanjavur	Villupuram	20	0.060	-
	-	-	-	-
	Total	100	0.300	102.5
M/s.Bannari Amman Sugars	Erode	75	0.225	150
Ltd., Erode	Total	75	0.225	150
M/s.Shree Harichandana				
Estates Pvt. Ltd				
Madurai	The firm has no programme for 2004-05.			
Theni				
M/s.Thiru Arooran Sugars Ltd.,	Perambalur	30	0.090	16.65
	Thanjavur	25	0.075	-
	Pudukottai	10	0.030	-
	Thiruvarur	25	0.075	-
	Nagapatti nam	10	0.030	-
	Total	100	0.300	16.65
	GRAND TOTAL	1475	4.425	324.65

For the scheme Government provides the firms exemption from Purchase Tax on Sweet Sorghum, Sales Tax on sale of Ethanol made from Sweet Sorghum at 8%. Special incentives as per industrial policy statement 2003 of Govt. of Tamil Nadu.

An extent of 1475 acres. was programmed to be covered by the above 6 firms, against which an area of 324.65 Acres was so far covered. The balance area is proposed to be covered by March 2005.

PROMOTION OF SUGARBEET

The firms M/s.Mohan Breweries amd Distelleries Limited, M/s. Bannari Amman Sugars Ltd., M/s. Thiru Arooran Sugars Ltd., M/s. Shree Harichandana Estates Pvt, Ltd., have furnished to Government their expression of interest for entering into contract farming agreements with farmers in selected districts, who will cultivate Sugarbeet and sell Sugarbeet tubers to the Processing Unit of the firms concerned and Government have permitted them to take up the programme in selected districts as follows. The benefits extended by Government under Sweet Sorghum on Purchase Tax, Sales Tax on ethanol and special incentives as per industrial policy statement 2003 of Govt. of Tamil Nadu are applicable for Sugar beet also.

In respect of sugar beet. it requires certain costly machineries like slicer, diffuser etc., which are not installed in the sugar factories at present. Further, the seeds are to be obtained only by import from other countries through seed companies.

Hence the firms are taking up the programme during 2004-05 only on trial basis. The programme will be extended for larger areas after studying the performance of the crop in the trials conducted during 2004-05 and after installation of required infrastructure facilities like slicer, diffuser etc., by the sugar industries.

Two firms have furnished their programme for 2004-05 to cover 160 Acres of which 17 Acres were covered. The balance 143 acres are proposed to be covered by the firms during March 2005. The promotional scheme for sweet sorghum and sugar beet will be continued during 2005-06.

Firm Name / Districts permitted by Govt.	Districts Programmed	Action Plan (acres)	Seed requirement (tonnes)	Area covered upto Feb 05 (acres)
M/s.Mohan Breweries & Distilleries Ltd Villupuram	Villupuram	50	60	0
	Tiruvallur	100	120	10
	Total	150	180	10
M/s.Bannari Amman Sugars Ltd., / Erode	Erode	10	12	7
	Total	10	12	7
M/s.Shree Harichandana Estates Pvt.Ltd	Madurai	The firm has no programme for 2004-05		
	Theni			
M/s.Thiru Arooran Sugars Ltd., Pudukottai, Thanjavur, Cuddalore Perambalur, Salem, Tiruvannamalai		The firm has no programme for 2004-05		
	GRAND TOTAL	160	192	17

SUGARCANE

India is the second largest producer of sugar next only to Brazil in the world and contribute around 1/5th of the World's Sugar Production. Tamil Nadu ranks first in productivity of sugarcane for the past several years. Sugarcane is cultivated normally in an area of about 3 L.Ha. which constitutes about 2% of total cultivated area of the State. The average annual production of cane is 320 L.MT. The productivity of sugarcane ranges from 92.9 MT/ha. to 124.2 MT/Ha. among the districts and the average State productivity is 107 MT/ha. The scope for increasing area under sugarcane is very limited as the crop is water intensive and hence focus is mainly to sustain the production and productivity of sugarcane in the State. The following activities are carried out through Sugarcane Development Scheme in all the districts except Ramanathapuram, Thoothukudi, Kanyakumari and The Nilgiris districts.

Sl. No.	Component	Target 2004-05	Achmt. upto February 2005	Programme for 05-06
1.	Release of parasite (Ha.)	10500	10250	10500
2.	Financial (L.Rs.)	3.75	3.430	3.75

Public Sector and Cooperative Sugar Mills

Cane Cultivation and productivity for 2005-2006 of Cooperative and Public Sector Sugar Mills. The Cooperative and Public Sector Sugar Mills have proposed to register cane area of about 77600 hectares for the 2005-2006 crushing season and expected to crush about 57.50 lakh tonnes as against 32.80 lakh tonnes planned to be crushed during the current year 2004-2005 crushing season with a capacity utilization of 50%.

Sugarcane Productivity of all Sugar mills in Tamil Nadu for 2005-2006

As the crushing operation in all the sugar mills are continuing, as on 28.2.2005, the mills have crushed 49.90 lakh MTs and at the end of the season, all the sugar mill in Tamil Nadu are expected to crush about 95.41 lakh tonnes. The average expected productivity of sugarcane is about 79 Metric Tonnes per hectare by adopting the yield improvement techniques. However, there is a scope to increase the productivity to about 100 tonnes per hectare by adopting the cane yield improvement techniques. The Government have approved an Action Plan to increase the yield in cane areas where it is less than the State average yield per hectare with the technical assistance rendered by the Sugarcane Breeding Institute and Sugarcane Research Stations.

Action Plan to increase Sugar Recovery

An action plan has been prepared to crush 57.50 lakh MTs with a capacity utilization of 88%, the expected average yield of about 74 tonnes per hectare with a sugar recovery of 10% during 2005-2006 crushing season as against 32.80 lakh MTs of cane arranged for crushing during 2004-2005 crushing season with an average recovery of 9.97 % and 69 tonnes average yield per hectare. To achieve the above objective it has been programmed to adopt the following strategies at field level.

1. Improving the cane productivity and area under High Sugar recovery rich Varieties.
2. Training to the farmers and field staff by the Scientist of Sugarcane Research Station, to impart the new technology.
3. Arrangement for timely supply of inputs and advice the cane growers to apply ferlisers in time.
4. Adoption of drought management techniques.
5. Promotion of Drip irrigation
6. Application of Bio-fertilizers and Bio-pesticides
7. Application of Bio-compost, vermi-compost and enriched press mud.
8. To create awareness among farmers and field staff by way of furnishing the integrated pest and disease management to control the net break of new pest and diseases.
9. Multiplication of high yielding new varieties.

Sugar Recovery and High Sugar Varieties

The high sugar varieties play a very important role in improving the sugar recovery of the mills and thereby reducing the cost of production. During 2004-2005 crushing season, the high sugar varieties to be crushed by Coop. Sugar Mills and

Public Sector Sugar Mills is about 90 %, with expected recovery of 9.97%. For the next season 2005-2006 also the high sugar varieties expected to be crushed with expected recovery of 10 %. Apart from this, the quantity of cane expected to be crushed for the next season would be 57.50 lakh tonnes as compared to current season's 32.80 lakh tonnes.

Promotion of Drip Irrigation for Cane Cultivation

Water is becoming a scarce resource and conventional irrigation methods for sugarcane involves excessive use of water and is highly labour intensive. The Government have planned to promote drip irrigation for sugarcane cultivation to conserve water and irrigate more areas. It has also been proposed to provide subsidy from Sugarcane Cess Fund to the farmers for installing drip irrigation system. For the above purpose, the Cess Fund Utilisation Committee have recommended to allot Rs.10/- lakhs from Cess Fund which is under consideration of the Government.

Application of Bio-Fertilizer and Bio-pesticides

Bio-fertilizers like Acetobactor and Phosphobacterium are applied in the soil for fixing atmospheric nitrogen and phosphorus respectively. Application of bio-fertilizers will reduce the requirement of chemical fertilizer and thereby minimize the cost of cultivation. Bio-fertilizer units will be started in Coop. Sugar Mills and Public Sector Sugar Mills to ensure sufficient supply of Bio-fertilizers.

Steps will be taken to supply Bio-pesticides like Trichogramma and NPV in all the Public Sector Sugar Mills. Application of Bio-pesticides will make sugarcane farming economically viable and environmentally safe. Bio-control Research Laboratory at Tamilnadu Agricultural University will be strengthened to ensure sufficient supply of Bio-pesticides.

Bio-compost

The total production of Press mud is about 2.00 lakh tonnes in all Cooperative and Public Sector Sugar Mills. Bio-compost is produced from press mud utilising the "spent wash" from the distilleries. At present there are only two distilleries-

Salem and Amaravathy Cooperative Sugar Mills where the entire press mud of about 25,000 tonnes are converted into Bio-compost (i.e. 11,000 tonnes of bio compost) and supplied to the ryots.

Vermi Compost

In other Cooperative and Public Sector Sugar Mills where distilleries are not available vermi compost is prepared by utilizing press mud and earthworms. Studies undertaken in Tamilnadu Agricultural University showed that Vermi-compost has positive impact on productivity of sugarcane and also other crops. Vermi-compost improves the soil texture and water holding capacity of the soil. Tiruttani, Cheyyar and Chengalrayan Cooperative Sugar Mills have taken up the scheme successfully. It is proposed to extend the scheme to other mills during the current year. Further, press mud is de-composed and mixed with bio-fertilizer and is distributed to the farmers as enriched press mud. All the sugar mills will be taking up this scheme and distribute enriched press mud to farmers to increase the yield and quality of cane. In the same manner cane trash is mixed with press mud and trash compost is made in the cane farm of the mills.

Cane Price Fixation

The Government of India in their notification announced Statutory Minimum Price for 2004-05 season at Rs.745/- per MT of cane linked to 8.5% sugar recovery, subject to a premium of Rs.8.80 per tonne for every 0.1% increase in the recovery above that level.

Collection of Cess and Sugarcane Road Development Scheme

The Sugarcane Road Development Scheme is being implemented by State Government to improve the network of link roads between the cane areas and the concerned sugar mills. The road network aims at facilitating easy movement and regular supply of sugarcane to the sugar mills. The funds for the Sugarcane Road Development Scheme are generated by levy of cess on sugarcane crushed by the sugar factories at the rate of Rs.5/- per tonne of cane purchased from outside the local area declared under the provisions of the Tamilnadu Sugar Factories Control Act, 1949. The details of cane cess collected during the last four years and expenditure on road works are given below

Year	Sugarcane cess collected (Rs. in lakhs)	Total Expenditure (Rs. in lakhs)
2001-2002	872.94	339.99
2002-2003	613.00	704.33
2003-2004	470.70	823.00
2004 - 2005 upto 28.2.2005	276.08	577.75

EXTENSION ACTIVITIES

Modernisation Of Extension System of Agriculture Department

The present system of extension is multi-layered and cumbersome. To have fast delivery system to the farmers, it is proposed to have a two tier system at the gross root level. Bearing this in mind, by providing all the infrastructure at the block level / village level, the extension wing can reach the farmers' door step. This will certainly help the farmer in receiving all the benefits from Agriculture Department which in turn certainly increase the production and productivity of the crops. The system which we are going to introduce will have a multi sectoral approach by coordinating all the service and allied sector.

Establishment of Agricultural Facilitation Centres

It is proposed to establish 200 Agricultural Facilitation Centres in 200 non taluk headquarters blocks at a cost of Rs.100 crores through RIDF loan subject to approval by NABARD. The Agricultural Facilitation Centres will have Consultancy Room of all Departments like Agriculture,

Horticulture, Animal Husbandry, Soil Conservation with Mini Laboratories , Exhibition centres and also Input godown. Providing Single Window Extension Service, bringing services closer to user and improved input supply are the major benefits of Agricultural Facilitation Centres.

Tamilnadu Women in Agri Business And Extension (TANWABE)

It is proposed to strengthen 4000 women group in 3 years and to make them as entrepreneurs in agricultural extension activities like raising seed farm, production of seedlings, hiring of agricultural implements and also to motivate the groups to involve themselves in various programmes.

Training Policy

Training has become an essential component in order to keep a competitive edge in the changing economic environment characterized by globalisation. Under this concept, it is aimed to identify the training needs and to conduct training on subjects related to farming technology which provide remuneration to the participants.

ICT Enabled Agriculture

As agriculture is the most vital sector of Indian economy, the ICT enabled Agriculture envisages the many facets of the agricultural sector viz., agricultural schemes, agricultural population, rainfall, land holdings, land use pattern, irrigation, cropping intensity, crop production technologies, productivity trend, agro climatic zonal planning, agricultural commodities prices, agricultural wages and implements etc. The State Agricultural Department is responsible to provide information to the farmers on crop production technologies, assistance through various development programmes, availability of inputs and its price, market trend, etc. for sustainable agricultural development.

Under the TN-AGRISNET project it is envisaged to equip the State and District level agricultural departments with the State-of-the-art ICT infrastructures, which will facilitate agricultural related information access and advisory services to the Resource-poor-farmers in particular, and all stakeholders of the supply-chain of the Agricultural system, in general. This will ensure that the computer systems are used not only for administrative purposes but also for "farmer-centric" Information System and Access. The programme is proposed to be implemented in two major phases.

1. Thrust will be given to empower the officers and staff of Agriculture department at every level under ICT.
2. Thrust will be to provide online services to the farming community and the external agencies involved in allied services.

In the first phase, all offices up to the Block level will have to be provided with adequate computing and communication infrastructure in order to capture, process and transmit the relevant details for further processing and dissemination. The applications will have to be primarily web based for easy implementation and maintenance. Good connectivity will have to be established between the offices. While planning for wide area connectivity, the availability of TN State Wide Area Network (TNSWAN) will have to be taken into account.

The TN-AGRISNET Server Nodes are required to be established at the Commissioner Office, 28 Joint Director of Agriculture Offices at Districts, 384 Block Agriculture Development Offices, 35 Soil testing labs, 9 Pesticide Testing Labs, 6 Fertilizer Control Labs, 6 Bio-Fertilizer Production units, 12 Farmers Training Centres, STAMIN, 37 State Seed Farms, 5 State Oilseed farms and 1 Pulses farm. This ICT activities in agriculture is proposed to be taken up with RIDF to the tune of Rs.19.06 crores loan subject to approval of the project by NABARD. Besides, an amount of Rs.5 lakhs is provided under Part II scheme to augment the ICT facilities to farmers linking (Rural Access Service through Internet) RASI Project during 2005-06 .

Web Based Data Collection for the Department

To ensure effective monitoring of the crop coverage, scheme progress and also administrative aspects, Monday Message Monitoring System on weekly basis, Personnel Information

System and File Monitoring System have been established at Commissionerate of Agriculture. This Web based Data collection helps to speed up the activities of the Department besides creating a Data Bank for future planning.

State Agricultural Extension Management Institute (STAMIN)

To train the extension personnel of the Department of Agriculture and to equip them with upto date technologies, approaches and also to enhance their managerial ability, a State Level Training Institute (STAMIN) is functioning at Kudumianmalai. The STAMIN is also engaged in conducting special training courses to NGOs, Agricultural Students, personnel from Sister Departments. The details of training courses conducted during 2004-05 are given below.

Sl. No.	Name of the Training	No. of Officers trained	
		Target	Achievement upto Feb. 05
1.	Office Administration Training	200	140
2.	Office Management Training	220	160
3.	Computer Training	100	80
	TOTAL	520	380

Farmers Training Centres

Dissemination of latest technologies to the farmers and ensure adoption is the prime responsibility of the Department. Training of farmers is a continuous process due to the change in agricultural scenario like cropping pattern, market preference, demand for agricultural produce, value addition, crop diversification as these are deciding factors to get profit in Agriculture. With the objective 12 FTCs are functioning in the state. These are located at Kancheepuram, Tindivanam, Vellore (originally functioned at Navlock), Salem, Dharmapuri, Erode (originally functioned at Bhavani) Lalgudi, Kudumianmalai, Sakkottai, Paramakudi, Palayamkottai and Nagercoil (originally functioned at Tirupathisaram). Besides imparting training at village levels, training cum demonstration, method demonstration and farmers tours are also conducted in collaboration with TamilNadu Agricultural University. Farmers day is also celebrated wherein lot of Farmers participate and share their experience.

Extension activities like farm news, through Radio, Newspaper and News letter are also carried out, to ensure speedy transfer of technologies. Target and achievement under Farmers Training Centre:

Sl. No	Particulars	Unit	2004-05		2005-06
			Target	Achmt. upto Feb 2005	Programme
1.	Village based training for men & women	Nos.	264	256	264
2.	Village based training for FDG conveners	Nos.	120	117	120
3.	Field demonstration	Nos.	1320	1266	1320
4.	Method demonstration	Nos.	528	510	528
5.	Financial Achievement (L.Rs.)		6.57	4.995	6.57

Considering the importance in training the farmers on latest technologies and to create opportunities for the farmers of all districts to get training benefits, the Hon'ble Chief Minister has ordered to establish new Farmers Training Centres in 11 districts. The 10 FTCs to be established at Thiruvannamalai, Namakkal, Karur, Perambalur, Dindigul, Theni, Sivaganga, Virudhunagar, Thoothukudi and Krishnagiri districts will functrion under Agriculture Department and one FTC at Nilgiris under the control of Horticulure Department during 2005-06 where the Farmers Training Centres belongs to the Department and Krishi Vigyan Kendras belongs to Tamilnadu Agricultural University are not functioning now. The 11 new Farmers Training Centres will be established during 2005-06 at a total cost of Rs.100 lakhs.

STUDY ON EXPORT PROMOTION

The WTO Agreement poses threats and opportunities. It is imperative that a detailed study and consultancy services is essential on Crop competitiveness both domestically and internationally and also the constraints in both forward and backward linkages to device modalities/ strategies to provide relevant recommendations to farmers and for proper implementation.

Under Ministry of Commerce, Government of India, a Market Access initiative (MAI) scheme is in operation wherein assistance are extended to supplement the State Government in carrying out export potential surveys of the States for identified product groups to evolve market related strategy for promoting exports of the identified product groups from the State. 50% of the cost of Survey will be provided as GOI assistance. Hence this study will be taken up on sharing basis at a cost of Rs.20.00 lakhs. (State Rs.10.00 lakhs + GOI (MAI scheme) Rs.10.00 lakhs) during 2005-06.

Crop Yield Competition

Crop Yield Competitions are conducted at the State level as well as at the District level to motivate the farmers to achieve highest productivity by adopting the advanced scientific techniques in crops like paddy, groundnut, cholam, cumbu, greengram and blackgram and cash prizes are awarded. The prize amounts given are as follows:

Crop	State Level (In Rs.)		District Level (In Rs.)	
	1st place	2nd place	1st place	2nd place
Paddy and groundnut	25000	15000	8000	4000
Other crops	8000	4000	4000	2000

The enrolment fee for paddy and groundnut crops for State level competition is Rs.100 and for other crops is Rs. 50 /-. Similarly, the enrolment fee for paddy and groundnut crops for District level competitions is Rs.50/- and for other crops it is Rs. 25/-. This programme is being implemented in 2004-2005 at a cost of Rs15.26 lakhs. This programme will be continued during 2005-2006.

CENTRALLY SPONSORED SCHEMES

Schemes Under Macro Management

The Government of India have launched the Macro Management Mode schemes in October 2000. This Macro Management Mode Scheme is a major step towards decentralization/ restoring primary role of the State in agricultural development planning.

Under Macro Management Mode, 27 identified Centrally sponsored Schemes were amalgamated and the States have the flexibility to develop programmes based on regional priorities through work plan comprising integrated approach on activities of agriculture, horticulture, Land Development, Soil Conservation, Marketing, Cooperation and State Land Use Board.

The expenditure under this scheme is shared by Government of India and State on 90:10 basis. Out of 90% of the GOI share, 80% is extended as grant-in-aid and balance 20% is treated as loan.

Under Macro Management Mode Scheme the following schemes are implemented by Agriculture Department.

Sugarcane Development Scheme under Work Plan

Government of India has approved inclusion of SUBACS (Sustainable Development of Sugarcane Based Cropping System) under Macro Management Mode Scheme. The scheme is implemented in all the districts except Ramanathapuram, Thoothukudi, The Nilgiris and Kanyakumari. Training of farmers on improved cultivation technologies and release of parasite for control of sugarcane borer is contemplated. Drip irrigation part of the scheme is implemented by the Agricultural Engineering Department. The progress under this scheme is as follows:

Sl. No.	Component	Unit	Physical		Financial (L.Rs.)	
			Target 2004-05	Achmt. upto Feb 2005	Target 04-05	Achmt. upto Feb 2005
1.	Farmers Training assistance @ Rs.5000/ training.	Nos.	100	97	5.0	4.850
2.	Release of Parasite @ 25% limited to Rs.10/Ha.	Ha.	5000	4634	0.50	0.463
	TOTAL				5.50	5.313

This scheme will be continued during 2005-06.

Balanced and Integrated Use of Fertilisers

The programme on Balanced and Integrated Use of Fertiliser is implemented under Macro Management Mode Scheme. The activities taken up under this programme during 2004-05 are:

Conducting Orientation Courses to Technical Personnel

Training is extended to 360 technical Officers on Micro Nutrient Analysis, uses of biofertilisers, organic farming, reclamation of problem soils and their management, etc. at Central Control Laboratory, Kudumianmalai in 12 batches at 30 Officers per training at a cost of Rs.1.20 lakhs during 2004-05.

Distribution of Soil Health Cards

The Soil Health Card is a semi permanent record containing details of recommendations on fertilisers, organic manure to be applied for the prescribed crop, besides recommendation on water quality, information on soil problem and relevant reclamation measures. During 2004-05, one lakh nos.of Soil Health Cards are being distributed at a cost of Rs.4.50 lakhs.

Distribution of Leaf Colour Chart

36,200 nos. of leaf colour charts is being distributed to the farmers with the leaflet containing the recommendation. The farmers can compare the chart with the leaf colour of the standing crop in their field to know the 'Nitrogen' need of crop and can apply the recommended quantity. This leaf colour chart will helps the farmers to apply optimum dose of 'N' by assessing the actual need of the crop and to avoid excess application of 'N' to the crop to reduce unproductive expenditure on fertilisers.

Printing of Soil Atlas

Soil and land based agricultural planning has been gaining momentum for increasing the agricultural production. Soil is the basic non renewable natural resource. Hence, printing of soil atlas for the left over 5 district of Villupuram, Cuddalore, Thiruvannamalai, Virudhunagar and Thoothukudi has been taken up at a cost of Rs. 4.8 lakhs during 2004-05. This Soil Atlas will contain various information like geology, geomorphology, land use pattern, land capability, land / crop suitability, soil depth, colour, slope, drainage, etc. The Soil Atlas provides holistic information at Taluk/ District level for Planner, Researcher, Development Officer, Field Functionaries and to a certain exigent directly to the farmers also.

Centrally Sponsored Scheme on Intensive Cotton Development programme under Mini Mission II of Technology Mission on Cotton

Technology Mission on Cotton was launched in February 2000 and become operartive from 2000-01 under Mini Mission II in Tamil Nadu with a aim of increase the production of cotton by providing key inputs, Demonstrations, Trainings and effective Plant Protection activities through Integrated Pest Management approach. The scheme is shared between central and state on 75:25 basis and implemented in all the districts of the State except Nilgiris, Kanyakumari, Thiruvallur and Kancheepuram.

During 2004-05, an amount of Rs.297.50 lakhs have been sanctioned for the implementation of the programme and further an additional amount of Rs.152.55 lakhs have been approved by Government of India totalling the scheme outlay to Rs.450.05 lakhs.

8

TRENDS IN HORTICULTURE GROWTH

Introduction

Tamil Nadu is one of the leading horticulture States in India contributing 7.7 per cent to the National Horticultural production with 5.7 per cent of the national level area. Tamil Nadu has been blessed with diversified agro-climatic conditions, suitable for a wide range of horticulture crops like fruits, vegetables, spices, plantation crops, flowers and medicinal plants. A large extent of wastelands and under-utilized lands are available in the State for horticulture development. Tamil Nadu has a long coastal belt of 1000 km. suitable for crops like cashew, coconut, tropical orchids etc. The southern part of Tamil Nadu has the potential for growing off-season mangoes and grapes. A lot of awareness has been created among the farmers of Tamil Nadu about cultivation of high value horticulture crops. It is aimed to achieve 8 per cent annual growth rate during X Five-Year Plan in the horticultural sector.

The area coverage in 2003-04 was 8.24 lakh hectares, a little less than the normal area of 8.52 lakh hectares. During 2004-05, despite the failure of the monsoon, the area coverage would still be around 8.91lakh hectares. In the year 2005-06, an area of 9.73 lakh hectares is expected to be covered.

TAMIL NADU HORTICULTURE DEVELOPMENT MISSION

The Government have setup a Mission for Horticulture development in Tamil Nadu vide G.O. Ms. No.155, Agriculture (H1-2) Department, dated 28.4.2003 with a mandate to give impetus to production, processing for value addition and marketing of Vegetables, Fruits, Flowers and Medicinal Plants. Tamil Nadu is the first state in India to setup a separate Mission for development of Horticulture in the year 2003. It is aimed to double the production of horticulture crops by the year 2011-12 in Tamil Nadu. The present level of productin of 99.47 L.MT will be doubled to 187.80 L.MTs during the year 2011-12.

Objectives of the Mission

1. Improving production through balanced nutrition management.
2. Evolving suitable mechanism for regulating the production of quality planting materials and giving impetus to need based research.
3. Establishing adequate infrastructure for post harvest management especially preservation and marketing.
4. Encouraging active involvement of farmers associations in adoption of modern technological practices.

A Governing Body has been constituted to monitor the effective implementation of the mission under the Chairmanship of the Honourable Minister for Agriculture with the Vice-Chancellor of Tamil Nadu Agricultural University, Secretaries to Government and Senior officers of various departments and Nominees of Central Government agencies like National Horticulture Board, APEDA (Agricultural and Processed Food Products Export Development Authority) as members. The Chief Secretary to Government is the Vice Chairperson of the Governing body.

Mango Development Mission

Under Mango Development Mission it has been programmed to layout demonstration plots in the farmer's field to enable them to have a visual impact of various new technologies involved in the cultivation of Mango crop. A sum of Rs.22.55 lakhs has been proposed for this purpose. National Horticulture Board has agreed to sanction Rs. 10.00 lakhs out of which released a sum of Rs.5.00 lakh for implementation of this scheme. On receipt of Government Order, this scheme will be implemented in the selected Mango growing districts.

Cashew Development Mission

Under Cashew Development Mission it has been programmed to layout demonstration plots in the farmer field to increase the productivity of cashew with high yielding varieties and hi technologies. A sum of Rs.23.00 lakhs has been proposed for this scheme. National Horticulture Board has agreed to sanction Rs.10.00 lakhs out of which released a sum of Rs. 5.00 lakh for implementation of this scheme. On receipt of Government Order this scheme will be implemented in the selected cashew growing districts.

Training to Farmers

With a view to expose the farmers to various hitech aspects of Horticulture, farmers are taken to various research stations in out of state on tour-cum-training basis. About 10,000 Nos. of farmers are given training during the year 2004-2005. Apart from this farmers were given awareness in various horticulture technologies of Medicinal Plants, Flower cultivation and Vegetable Cultivation through the conduct of District level seminars and Exhibition. The extension officers and field level workers of this department are also given training in the latest technologies of horticulture from the Research Stations situated inside and outside the State.

STATE PLAN SCHEMES

The following State Plan Schemes have been implemented during 2004-05.

Integrated Horticulture Development Scheme (IHDS)

In G.O. Ms. No.180, Agriculture (H2) Department, dated 28.5.2004, Integrated Horticulture Development Scheme was sanctioned with an outlay of Rs.766.26 lakhs. Under this scheme, planting materials, high yielding/hybrid vegetable seeds are distributed to horticultural farmers at 50% subsidised cost.

Integrated Tribal Development Programme (ITDP)

In G.O. Ms. No.189, Agriculture (H2) Department, dated 28.7.2004, Integrated Tribal Development Programme was sanctioned with an outlay of Rs.39.325 lakhs. Under this scheme, horticulture inputs are distributed to Tribal farmers in Salem, Namakkal, Dharmapuri, Tiruvannamalai, Vellore, Trichy and Villupuram districts at subsidised cost. Tour-cum-Training is also given to the Tribal farmers under this scheme.

Western Ghats Development Programme (WGDP)

In G.O. Ms. No.52, Plannting, Development and Special Initiatives (TC-1) Department, dated 22.7.2004, Western Ghat Development Programme was sanctioned. Under this scheme, horticulture inputs are distributed to the horticultural farmers in Theni, Madurai, Dindigul, Virudhunagar, Coimbatore, Erode, Tirunelveli and Kanyakumari districts at 25% subsidised cost.

Hill Area Development Programme (HADP)

In G.O. Ms. No.60, Planning, Development and Special Initiatives (TC 1 & 2) Department, dated 5.8.2004, Hill Area Development Programme was sanctioned with an outlay of Rs.315.00 lakhs. Under this scheme, horticultural inputs are distributed to the farmers of Nilgiris district at 25% subsidised cost.

The above Schemes will be continued during the year 2005-06 also. In the above schemes:-

- Focus is given for cultivation of high yielding / hybrids and varieties, adoption of efficient water and nutrient management and Integrated Pest Management.
- Adoption of location specific and crop specific technologies is encouraged.
- Mango, Amla, Guava, Sapota and Vanilla are given importance under fruits and spices categories. High yielding / export potential varieties are given priority.
- Under vegetables, focus is given on hybrids / high yielding varieties which fetch good return to the farming community.
- Under Integrated Horticulture Development Scheme, horticulture plants are distributed at 50% subsidy for a maximum of 1 hectare for fruits and 0.5 hectare for seeds of vegetables and flowers to each beneficiary.

- In the Tribal areas of Tamil Nadu, viz., Shevroy hills, Arunoothumalai, Pachamalai, Kolli hills, Sitheri hills, Javvadu hills, Kalrayan hills, and the tribal farmers are encouraged to develop individual orchards by optimally utilizing the available water resources.
- Considering the economic status of the tribal farmers, the horticulture plants are distributed at 75% subsidy under the Integrated Tribal Development Programme (ITDP).
- Watershed approach is followed in the Western Ghat Development Programme and the Hill Area Development Programme.
- Focus is given on prevention of soil erosion and preserving the Eco-System through appropriate cropping pattern.
- Planting materials and seeds are distributed at 50% cost under Western Ghat Development Programme and Hill Area Development Programme.
- The State is maintaining 55 State Horticulture Farms (Annexure I) in which Pedigree and Hybrid planting materials are produced and supplied to the farmers under various schemes.
- To update the technical knowledge of the extension officers of the Horticulture Department, in-service training is given at Central Horticulture Training Centre, Kudumianmalai in Pudukkottai district.
- A two-year diploma course is being conducted at the Horticulture Training Centre, Madhavaram (Chennai) with an intake of 40 students every year.

Promotion of Alternate Cropping in the Nilgiris

Promoting alternate crops is considered as an important measure to overcome the crisis facing the tea growers. Steps are being taken to popularize multi-tier cropping, viz. cultivation of silver oak, pepper, cardamom, mandarin orange, etc., along with tea plantation. Considering the prevailing agro-climatic condition in The Nilgiris, new kinds of crops like Macadamia and Peacan nuts have been identified for introduction. It is proposed to raise these crops in the State Horticultural Farms at Nanjanad and Colegrain in Nilgiris district on trial basis. Subsequently, efforts would be made to introduce these crops to the farmers in select pockets. A sum of Rs.50.00 lakhs has been sanctioned vide G.O. Ms. No.244, Agriculture (H2) Department, dated 17.6.2004 by the Government for promoting alternate cropping in Nilgiris district and the scheme will be continued during 2005-06 also.

Tamil Nadu Precision Farming Project

As announced during last budget session, Government have sanctioned an Innovative Scheme by name "Tamil Nadu Precision Farming Project" to be implemented in Dharmapuri District and newly formed Krishnagiri district covering an area of 400 Ha. of farmers' land with Micro Irrigation and Fertigation vide G.O. Ms. No.24, Agriculture (H1) Department, dated 23.1.2004. This scheme will be implemented with a total cost of Rs.720.60 lakhs over a period of 3 years.

It is programmed to cover an area of 100 Ha. in the first year, 200 Ha. in the second year and 100 Ha. in the third year at financial allocation of Rs.208.59 lakhs, 316.16 lakhs and Rs.195.85 lakhs respectively.

The Tamil Nadu Agricultural University will undertake this project as a Turnkey project and implement it with the co-operation of the Departments of Horticulture, Agricultural Engineering, Agriculture, Agricultural Marketing and Agri-Business and the District Administration. High value like gherkins, hybrid tomatoes, capsicum, paprika, babycorn, white onion, bhendi, cabbage and cauliflower are proposed to be cultivated under the scheme. Under this project, 100% subsidy will be given on the cost of cultivation of the first crop to the farmers selected during first year. 10% of the cost of cultivation will be collected from the farmers selected during second year. 20% of the cost of cultivation will be collected from the farmers selected during third year.

During the year 2005-06, the above scheme will be implemented in six more districts viz., Theni, Vellore, Cuddalore, Erode, Tirunelveli and Thanjavur with an financial outlay of

Rs.10.00 crores. An area of 100 Ha. will be covered in each district.

PART II SCHEMES

Under Part II Scheme a sum of Rs.120.00 lakh was sanctioned vide G.O. Ms. No.398, Agriculture (H1) Department, dated 6.8.2004 to improve the irrigation facilities in 28 State Horticulture Farms and the work is entrusted with Public Work Department and Agriculture Engineering Department and will be completed before 31.3.2005.

Tamil Nadu Horticulture Development Agency (TANHODA)

An independent nodal agency viz., The Tamil Nadu Horticulture Development Agency (TANHODA) has been formed vide G.O. Ms. No.250, Agricultue (H1) Department, dated 18.6.2004 and Rs.20.00 lakhs was sanctioned under Part II 2004-05 with the following objectives.

1. To promote Hi-Tech Horticulture and Precision Farming.
2. To promote Public Private Partnership (PPP) for horticulture development.
3. To disseminate knowledge on modern technologies in horticulture to the farming community as well as extension personnel.
4. To manage production and distribution of pedigree planting materials through State Horticulture Farms.
5. To implement the horticulture development schemes including the schemes under the Tamil Nadu Horticulture Development Mission and the National Horticulture Mission by accessing funds from the agencies concerned.

This agency will serve as a special purpose vehicle (SPV) for the implementation of various horticultural development programmes.

Part II Schemes (2005-06)

The following new schemes will be implemented during the year 2005-06.

1. Scheme for provision of infrastructure facilities in 25 State Horticulture Farms (Farm equipments like Power Tillers with Accessories and Multipurpose Tanker) - Rs.50.00 lakhs.
2. Organic farming vegetable cultivation with market linkage in 15 State Horticulture Farms - Rs.10.00 lakhs.
6. Water Management - Command Area Development Programmes (Centrally sponsored scheme - shared between State and Centre)

Command Area Development Programme

The Command Area Development Programme is implemented by the Agricultural Engineering Department as a centrally sponsored and equally shared programme on 50: 50 basis between the State Government and Government of India. At present, the programme is implemented in Cauvery Command, Parambikulam Aliyar Project, Tambirabarani River Basin Project, Palar-Poranthalar Project, Krishnagiri Reservoir Project, Gadana Ramanadhi Irrigation System, Nambiyar River Basin System, Patchaiyar River Basin System, Manimuthar Irrigation System, Aanaimaduvu System, Chinnar Resevoir system and Maruthanadi System.

The programme has been completed in Lower Bhavani Project, Cumbum Valley Project, Periyar - Vaigai Project, Sathanur Reservoir Project, Amaravathy Reservoir Project, Kothayar-Chittar project and Thoppaiyar Project. The remaining projects will be covered as per the schedule approved by the Government of India.

The six major components of Command Area Development Programme are as follows:

a. Construction of Field Channel
b. construction of Field drain.
c. Reclamation of Water logged Saline and alkaline soil areas.
d. Rotational Water Supply.
e. Participatory Irrigation Management
f. Management Subsidy.

The Agricultural Engineering Department has also developed a water management strategy for the area below the sluice, which is crucial for agricultural production. Water management holds the key for optimisation of water use and maximisation of crop production.

Construction of Field Channel

Construction of Field channel is done by considering the needs of the head reach farmers so that their interference is eliminated and the tail end farmers also get adequate irrigation water. This brings in equity in distribution of water to all the farmers in the entire sluice command. This has been achieved by using optimum flow concept.

The unit cost is Rs.10000/- per ha. in all New Project areas including Cauvery Command. For old ayacut of Cauvery Command the unit cost of construction of field channel works is Rs.4400/- per ha. An area of 9.87 lakh ha. has been covered till 31.3.2004 out of total cultivable command area of 10.97 lakh ha. with Construction of field channel component in 12 command areas. During 2004-2005, this work is being carried out in an area of 25695 ha. It is programmed to continue the scheme during 2005-2006 also.

Construction of Field Drains

Under the construction of field drains component, drains from individual fields to Government drains (or) natural drains outside the outlet are executed. The expenditure on field drains is inclusive of costs of earthwork, road cuttings and drop structures, if needed to prevent erosion. The unit cost per ha. for this component is Rs.4000/-

Reclamation of Water Logged, Saline & Alkaline Soil Areas

In this component, the area affected by Waterlogging, saline and alkaline problems within the command area are identified and measures to reclaim those patches are takenup. At present this component is taken up in Tamirabarani River Basin Project and Krishnagiri Reservoir Project at an unit cost of Rs.15000/- per ha.

Rotational Water Supply Works

Rotational Water Supply or Warabandi is a system of equitable water distribution, by turns according to a predetermined schedule specifying the day, time and duration of supply to each farmer in proportion to his holding size in an outlet command.

The rotational water supply schedule is prepared by the staff of Agricultural Engineering Department after executing the on farm development works and handed over to farmers for implementation.

Rotational water supply was first introduced in Lower Bhavani Project, Periyar - Vaigai Project and Cauvery Command in the year 1984-'85. An area of 8.92 lakh ha. has been covered under rotational water supply in Tamil Nadu in 12command areas till March 2004. During 2004-2005, this work is being taken up in an area of 52750 ha. It is programmed to continue the scheme during 2005-2006 also.

Participatory Irrigation Management

To implement On Farm Development Programmes and for ensuing equity in distribution of water, farmers participation is important.

Agricultural Engineering Department is intensively encouraging itself in this work for the last 20 years. The task of handing over the responsibilities to the farmers is being implemented. In orders to ensure farmers participation, farmers associations are formed on a three tier basis as follows.

Farmers' Association - at sluice command level

Farmers' Council - at distributory level

Farmers' Federation - at project level

Management Subsidy

In order to sustain the farmers associations/ councils already formed and for the maintenance of the on-farm infrastructures created and to continue RWS (Warabandhi) for equitable water distribution, a financial assistance is provided in the form of one time functional grant to registered Farmers' Councils at Rs.540.00 per ha. shared equally by the Central and the State Governments. To avail this, farmers have to contribute Rs.60.00 per ha.

Out of the 632 Farmer's Councils registered, an amount of Rs.1478.95 lakhs has been released as management subsidy till March 2004 to 597 registered Farmers' Councils to maintain the infrastructure created under the Command Area Development Programme.

World Bank Aided Model Rehabilitation Project in Hanumanadhi Sub Basin

The World Bank aided IAIP pilot project was implemented in Hanumanadhi sub basin covering 135 ha. in the command area, which is fed by Adavinainarkoil system tanks, at a cost of Rs.81.00 lakhs. This project is construed as a threshold for a broader Model Rehabilitation Project in Hanumanadhi sub basin. The World Bank aided model Rehabilitation Project in Hanumanadhi sub basin of Tamirabarani river basin in Tirunelveli district has been successfully completed by September 30th 2004.

The project has been executed and implemented in a multi disciplinary project approach involving PWD-WRO, Departments of Agricultural Engineering, Agriculture, Horticulture, Fisheries, Agricultural Marketing and TNAU. The funds of Rs.62.5 Crores was provided vide GO MS No. 33 PW (WR 1), Dated 30-1-2004, out of which Rs. 22.55 Crores was allotted to Agricultural Engineering Department. The project area encompasses Sengottai, Tenkasi and V.K. Pudur taluks.

The ayacut irrigated by 50 tanks and 14 anicuts is spread over an area of 3000 ha. in 15 villages namely Ayikudi, Elathur, Kilankadu, Kodikurichi, Kulayaneri, Kutthukalvalasai, Nainaragaram, Neduvayal, Panpoli, Rajagopalaperi, Sambavar Vadagarai, Surandai, Sundarapandiyapuram, Verakeralampudur and Vadagaraikilpidagai..An area of 2194 Ha. has been covered with an expenditure of Rs. 16.35 crores by the Agricultural Engineering Department.

Objectives

- Improving productivity of water and crop production by sharing of water and Mechanized farming.

- Uniform and optimum application of water and fertilizer by high tech micro irrigation systems. Rain water harvesting and recharge of ground water through rejuvenation of wells and farm ponds.
- Training, effecting attitudinal changes among the stakeholders and improving their socio - economic status with measures like Aquaculture in farm ponds,

High Tech Drip and Sprinkler facility

In the pre project period only two crops were cultivated in a year and most of lands were kept fallow for want of water in the third season After implementation of the project it is made possible to cultivate an additional third crop using the water conserved through micro irrigation techniques. After the project implementation an intensive cropping scheme one like below is followed.

I. Crop : 28 Ac-inch water required crops like groundnut, vegetables.

II. Crop : Paddy.

III. Crop: 18 Ac-inch water requirement crops like Pulses, fodder cholam etc.

The introduction of drip and sprinkler irrigation systems in the project area on a 10 Ha block basis is estimated to save atleast 30 - 40 % total irrigation water. Sprinkler irrigation system are provided to the individual farmers who do not have wells and new borewells are sunk for these farmers. Drip irrigation systems are provided to the farmers who have atleast one well.

Under ground PVC pipes are laid to convey water to the micro irrigation systems from the sources of water. However open field channels shall carry water during the paddy season. An area of 2194 Ha., is covered with an expenditure of Rs.1065.72 crores under high tech micro irrigation system including pipe laying.

GIS and Survey

The geographical information survey is used to get precise information on land holding, water bodies, vegetation etc., The

GIS format creation in 1:1000 scale, high resolution satellite data like IRS PAN merged 1:10,000 scale, GIS in 3-D, aerial photographs are some of the advanced survey and map preparation techniques.

ISDN / Lease line connections are installed at offices of water users groups, AED office in Tenkasi, Tirunelveli and MDPPP office AED, Commissioner of Agricultural and Commissioner of Agricultural marketing, secretariat in Chennai to provide latest information on farm practices and market trends to the farmers. These facilities have been created with Rs. 2.00 crores.

Water Harvesting Structures

Farm ponds are executed for collection of rainwater. 52 ponds have been constructed.

The pond water is used for domestic, irrigation and aquaculture purposes. The defunct and low yielding wells are rejuvenated with vertical and horizontal in - well bores to improve their yield. Shallow and medium bore wells are sunk and the new bore wells are provided with energized motors. An amount of Rs. 2.84 crores has been incurred on these water-harvesting measures.

Farm Mechanization

Modern agricultural machineries like power tillers, paddy transplanters etc. have been procured at the cost of Rs.0.53 crores and are hired to the farmers.

Training to Farmers and Officers

Trainings on high tech micro irrigation systems, water-harvesting methods and farm mechanization are imparted to the water user groups and departmental officers with Rs. 0.34 crores. Documentation, video graphy and evaluation studies of the project are also included in the programme.

Farmers' Participation

The participation of the stakeholders - the farmers, is emphasized right from the decision-making and implementation stages. The farmers contributed in cash, kind and labour at various stages of the project. The progressive farmers, the women farmers and rural youth participated in every sphere of activity. The farmers will be sharing a portion of the project cost in the form of betterment levy.

Follow up on Project

With the funds from the ongoing schemes, state and other organizations like NABARD etc., the uncovered areas in this project are proposed to be covered.

JATROPHA

It has been proposed to taken up Jatropha Cultivation in 200 Ha. funded by NOVOD Board during the year 2004-05 in the following 11 State Horticulture Farms.

1 G.O. Vallathirakottai, Pudukottai District 45 Ha.

2 G.O. Karumanthurai, Salem District 50 Ha.

3 SHF, Jeenur, Krishnagiri District 20 Ha

4 SHF, Kudumianmalai, Pudukottai District 10 Ha

5 SHF, Navlock, Vellore District 20 Ha

6 SHF, Mulluvadi, Salem District 10 Ha

7 G.O. Srivilliputhur, Virudhunagar District 12 Ha.

8 SHF, Melkadirpur, Kancheepuram District 10 Ha

9 SHF, Vichanthangal, Kancheepuram District 10 Ha

10 SHF, Thagarakuppam, Vellore District 10 Ha

11 SHF, Kannampalayam, Coimbatore District 5 Ha

Total 200 ha.

NOVOD has sanctioned a sum of Rs.75.00 lakhs to the Managing Director, Tamil Nadu Watershed Development Agency, Chennai 32 towards implementation of this programme.

TARGETS AND ACHIEVEMENTS

The financial allocation for the year 2004-05 and the expenditure incurred upto February 2005 and the proposed outlay for the year 2005-2006 for the State Plan Schemes and Schemes shared between State and Centre are furnished below.

Allocation for the year 2004-2005 and Proposed outlay for 2005-2006.

Unit: Rs. in lakhs

Sl. No	Scheme	2004 – 2005 Amount Sanctioned	Acht. upto Feb. 05	2005-2006 Proposed Outlay
A. State Plan Schemes				
1	Integrated Horticulture Dev. Scheme	766.260	615.080	769.400
2	Integrated Tribal Dev. Programme	39.330	35.135	41.790
3	Western Ghat Dev. Programme	135.480	105.158	132.080
4	Hill Area Development Programme	315.000	267.750	314.940
5	Precision Farming in Dharmapuri Dt.	232.950	203.000	1120.850
6	Promotion of alternate cropping(Nilgiris Dt)	50.000	48.090	50.000
7	Part-II Schemes	120.000	105.000	60.010
8	TANHODA	20.000	0	20.000
	Total	1679.020	1379.213	2509.0 0
B. Schemes shared between State and Centre				
1	Integrated Prog. for Dev. of Cashew	155.200	141.847	273.320
2	Integrated Prog. for Dev. of Fruits	218.000	186.389	318.350
3	Integrated Prog. for Dev. of Spices	60.000	47.946	180.130
4	Integrated Prog. for Dev. of Vegetables	219.000	191.644	310.000
5	Integrated Prog. for Development of Medicinal and Aromatic Plants	52.500	48.350	52.700
6	Integrated Prog. for Dev. Cocoa	12.000	11.190	12.000
7	Dev. of Commercial Floriculture	145.820	111.200	165.000

8	Development of Mushroom cultivation	10.000	9.500	10.000
9	Development of Horticulture through plasticulture interventions	14.500	13.030	14.500
10	Innovative Programmes for Development of Horticulture	113.900	43.455	90.000
	Total	1000.920	804.471	1426.000
	Grand Total (A+B)	2679.940	2183.684	3935.070

PROGRAMME IMPLEMENTATION FOR 2005-06

- Thrust on Hi-Tech Horticulture and precision farming.
- Expansion of area under micro-irrigation and fertigation.
- Stabilizing the area of water loving crops and expanding the area under dry land crops with a focus on effective water management.
- Area expansion of horticulture crops through various Waste Land Development programmes.
- Strengthening the system of production of pedigree and hybrid planting materials in both public and private sectors.
- Promotion of Organic farming with focus on export market.
- Promotion of Agri Export Zones (AEZ) for specified crops.
- Building up of Public and Private Partnership.
- Promotion of Contract / Corporate farming.
- Empowerment of farmers with special focus on farm women.
- Effective transfer of technologies by tour-cum-training to farmers.
- E-Governance and Human Resources Development through effective training for extension officers.
- Linkage with Agro Processing Industries with New Anna Marumalarchi Thittam (NAMT).
- Post harvest management and reduction of post harvest losses.

9

AGRICULTURAL ENGINEERING

INTRODUCTION

The Government of Tamil Nadu with a vision on the future has launched a number of projects and programmes with appropriate technologies for sustainable land and water management.

The Agricultural Engineering Department (AED), is engaged in the Conservation, Development and Management of the agricultural land and water resources of the State. "Watershed Management" and "Water Management" are the two major strategies of the department in contributing to the sustainable increase in agricultural production.

Through integration of rain water harvesting with precision and high tech farming and monitoring through Geographical Information System, the Agricultural Engineering Department proposes to move towards saving every drop of rain water, ensuring more crops per unit of water and adequate supply of water for agriculture.

This department implements a number of State, Centrally sponsored and Externally aided schemes / projects with the following objectives:

- To prevent soil erosion, improve soil moisture and create water-harvesting facilities in the watersheds, predominantly under rain fed agriculture.
- To promote appropriate land use, arrest soil erosion and improve the health of River Valley Catchments, Western Ghats, hill areas, and tribal areas.

- To develop degraded wastelands with appropriate land use practices by using high technology like Geographic information system (G.I.S).
- To optimise water use and to ensure equity in water distribution and create farmers' organisations for sustainable water management in the Command areas.
- To create new irrigation facilities and stabilise the existing facilities by harnessing ground water where it is sustainable.
- To optimise water use by installation of Drip Irrigation and Sprinkler Irrigation Systems and to propagate precision and high-tech farming by integrated farming approach.
- To provide Farm Power to carryout timely agricultural operations right from land levelling, shaping, ploughing, harrowing and puddling to harvesting.
- To make existing pump sets energy efficient so that farming cost is reduced and energy is conserved.
- To promote Rainwater harvesting so that ground water is recharged.

SOIL AND WATER CONSERVATION PROGRAMMES

Soil and Water Conservation in the Catchments of River Valley Project

Sedimentation of reservoirs is a serious problem posing a threat to the life of the reservoir, its storage capacity, the command area and generation of power. Sedimentation is caused due to soil erosion and improper land use in the catchment areas. With a view to reduce siltation of the reservoirs, various soil conservation measures are taken up in the River Valley Catchments of Tamil Nadu. The Soil Conservation works are taken up on watershed basis in the Very High and High priority watersheds identified by the All India Soil and land Use Survey Organisation (AISLUS) and approved by Government of India depending upon the rainfall, nature of erosion, soil texture, slope and topography of the area. During 2004-05, the programme is being implemented at a cost of Rs.734.41 lakhs covering 5965 ha. in South Pennaiyar & Mettur Catchments in Dharmapuri and Krishnagiri districts. During 2005 - 06, it is proposed to cover 6200 ha. in South Pennaiyar & Mettur Catchments at a cost of Rs. 1000.00 lakhs.

Soil Conservation in Tribal Areas

The objective of the scheme is to improve the economic status of the tribal farmers so as to lift them above the poverty line. By conserving the soil, improving the water holding capacity and improving the land use, the land productivity is proposed to be increased. Integrated development of tribal areas has been taken up under a multi sector approach in Jawadhu hills (Vellore and Tiruvannamalai district), Kalrayan hills (Salem and Villupuram districts) Shevaroy hills (Salem district), Sitheri hills (Dharmapuri district), Kolli hills (Namakkal district) and Pachamalai (Trichy district). Land Levelling, construction of contour rubble bunds with vegetative hedges and check dams are taken up in tribal areas at Government cost. These programmes provide employment to the local people and also help them to bring the treated areas under cultivation. During 2004-2005, this scheme is being implemented to treat 699 ha. of tribal lands at a cost of Rs.92.90 lakhs. During 2005-06, it is proposed, to implement the scheme to treat 600 ha. of tribal lands at a cost of Rs.92.90 lakhs.

Soil and Water Conservation under Hill Area Development Programme

Soil conservation scheme is being implemented in the Nilgiris District as an integrated watershed development programme, benefiting all the farmers irrespective of category and crops. The Nilgiris district has been delineated into 75 watersheds. Among these watersheds, highly degraded watersheds are taken up for treatment. Besides, soil conservation works, landslide preventive measures etc., have been taken up. Under this scheme, 100% grant is provided for soil conservation measures. However, beneficiaries are required to contribute at the rate of 10% for individual works and 5% for community works. In the case of SC/ST communities, the contribution required is 5%. The landslide treatment measures are however, executed under 100% grant. During the year 2004 - 05, this programme is being implemented to treat an area of 649 ha. with an outlay of Rs.358.01 lakhs. During 2005-06, it is proposed to treat an area 520.500 ha. with the financial outlay of Rs. 560.310 lakhs.

Soil and Water Conservation under Western Ghat Development Programme

Soil Conservation is one of the important aspects under Western Ghats Development Programme. The works are taken up in the Western Ghats Region covering the districts of Coimbatore, Erode, Dindigul, Theni, Virudhunagar, Tirunelveli, and Kanyakumari. During 2004-2005, Soil Conservation works in 645 ha. are being taken up at a cost of Rs.327.36 lakhs in eight districts. During 2005-2006, an area of 741 ha. is proposed to be covered at a cost of Rs.347.754 lakhs in eight districts.

Soil and Water Conservation Scheme for Rain Water Harvesting through Farm Ponds and Rejuvenation of Failed / Unused Wells

Over the years the area under well irrigation is increasing, resulting in increased ground water exploitation. The increased exploitation of ground water in many areas of the State has resulted in lowering of water table below the economic pumping level. Mining of ground water has occurred due to sinking of bore wells up to 1000 feet in certain areas. The ground water potential has been exploited to such an extent that special methods of rainwater harvesting and ground water recharge are warranted to save the well irrigated areas. Government have provided resilience to the drought affected agriculture by promoting rain water harvesting. They have sanctioned the construction of farm ponds and rejuvenation of failed / unused / abandoned wells. A total of 8833 farm ponds were constructed at a total cost of Rs. 2564.29 lakhs and 2093 wells have been rejuvenated at a cost of Rs.264.37 lakhs during 2003-04. An outlay of Rs.10.00 crores has been provided in the BE 2005-06 for taking up soil and conservation works. The Agricultural Engineering Department will plan rain water harvesting and run-off management on mini watershed basis. This proramme will be designed in such a way that the watershed will get the benefit of convergence of various watershed development and individual beneficiary oriented programmes. A dynamic monitoring and evaluation methodology will be inbuilt in the programme. Rain water harvesting will be given further importance and will be continued in 2005-06 under all the Soil and Water Conservation programmes and NABARD assisted Rain Water Harvesting under RIDF Programme. Rain Water Harvesting structures such as farm ponds, check dams and rejuvenation of abandoned wells are proposed to be taken up during 2005-06, with the NABARD assistance under Rural Infrastructure Development Funds so as to benefit 10,000 farmers.

Pilot Project for Development of Coastal Saline and Sand dune Affected Soils of Nagapattinam - Vedaranyam Zone

The Coastal areas of Nagapattinam district Tamil Nadu consists of sand dunes and sandy belt with fresh water overlying saline ground water. Sea water intrusion is noticed in the Vedaranyam canal. The Coastal Zone in between Nagapattinam and Vedaranyam has natural resources, but have problems posed by the sea water intrusion as well as the climatic conditions. This pilot project aims at developing the lands affected by sand dunes, sandy plains with coastal aquifer, the saline belt and the water logged areas in a more scientific way. The three-year project is being implemented by this department in association with the Tamil Nadu Agricultural University (TNAU) Coimbatore and with farmers using a Participatory Rural Appraisal (PRA) approach. Government of India has accorded sanction for Rs.1.20 crores which includes the Government share of Rs.89.92 lakhs and beneficiary share of Rs,30.08 lakhs for this three year project to cover an area of 2000 ha. During 2003-2004, the programme was implemented in 8 villages at an area of 792 ha. for Rs.31.18 lakhs. During 2004- 2005, this programme is being executed with the full .assistance of Central Government at a cost of Rs 34.217 lakhs in an area of 810 ha. During 2005-2006, the final phase of this programme will be implemented to cover the balance area of 398 ha. at a cost of Rs.24.73 lakhs.

Centrally Sponsored Scheme on Reclamation of Alkali soils

The Centrally Sponsored Programme for Reclamation of Alkali Soils in the State aims at improving the productivity of the alkaline lands by adopting suitable package of alkali soil reclamation technology. An outlay of Rs.690.00 lakhs has been approved for reclamation of 5000 ha. of alkaline soil during the years 2003 - 2004 and 2004 - 2005. During 2003-2004, an area of 929.24 ha. have been treated at a cost of Rs.19.26 lakhs. During 2004-2005, this scheme is being implemented in 1527 ha. of alkaline soils with an outlay of Rs.50.00 lakhs. During 2005-2006, it is proposed to continue the programme to cover an area of 3054 ha. at a cost of Rs.100 lakhs.

REMOTE SENSING

Remote Sensing and Geographic Information Research Centre at the O/o. Chief Engineer, AED, Chennai

The Remote Sensing Centre of the Agricultural Engineering Department provides input for various Watershed Development Programmes and Wasteland Development Programmes. Under Western Ghats Development Programme, the status report for western ghats region are prepared annually on watershed basis depicting the Geology, Geomorphology, Infrastructure, Micro watershed codification and soil erosion for prioritisation of the watershed activities. This centre has been recognized by Anna University as Research Centre in which Ph.D, and M.S programmes of Anna University can be carried out by Department staff on external registration. This centre can also guide research programmes of Anna University. The State Land Use Board has given a study to prepare Watershed Atlases and Soil Watershed Atlases for all the districts in Tamil Nadu. Such maps have been used to delineate the micro water sheds under NWDPRA and Comprehensive Waste land development projects.

District GIS Centre

A Geographical Information System Centre has been created at Theni and Salem. This will help to build database of all the resources available at watershed level for taking up various development activities at micro watershed level using GIS Software. The Watershed Development Programmes like River Valley Project, IWDP, DPAP, NWDPRA, etc., are carried out by this department all over Tamil Nadu. It is proposed to have information on watersheds covered under various programmes on 1:10000 scales and create a GIS for each district. Maps will be created for each project area in which information upto the field level in the micro watersheds will be built including field boundary, SF No., land use, soil-cropping practices, etc. At the micro watershed level, runoff and soil erosion will be worked out and soil conservation works will be designed accordingly. Similarly, one GIS centre is proposed to be established in Trichy during 2005-06.

MICRO IRRIGATION

The Agricultural Engineering Department is engaged in popularising Micro Irrigation systems (Both Drip & Sprinkler) in Tamil Nadu, since the year 1985 - 86. Among the three sources of irrigation namely, canals, tanks and wells, the wells account for 51.70% of the net area irrigated. In order to reduce the stress on well irrigation and also to ensure judicious and economic use of irrigation water, through installation of Micro Irrigation Systems, viz. Drip / Sprinkler Irrigation Systems, an allocation of Rs 51.22 crores was made in 2004-2005 and the works are under progress. The scheme conceived by the State Government supports all the farmers for all the crops and will be continued. During 2005-2006, it is proposed to continue the programme with an outlay of Rs.17 crores. A plan has been laid for ' implementing Micro Irrigation system in State Seed Farms covering an area of 300 Ha during the year to popularise the Jetropha, Sweetswargham and Sugarbeet .

MINOR IRRIGATION

Tube Wells Scheme

Creation of additional irrigation facilities, stabilisation of the area already under irrigation and promotion of conjunctive use of surface water and ground water by harnessing groundwater resources are the objectives of the Tube Wells Scheme. The State of Tamil Nadu has nearly exhausted the surface water sources and there is increasingly more dependence on ground water. Therefore, sinking of new tubewells and revitalisation of dried up wells are taken up under this programme. Under this scheme, the Resistivity meters and Electrical loggers are provided to farmers for locating well sites and aquifers. Drilling Rigs, Mini Drills, Hand Boring Sets, Rock Blasting Units, Long Hole Equipment and Air Compressors are also provided to farmers on hire basis to take up these activities. The Department maintains a fleet of minor irrigation machinery of different types. During 2004-05, the programme is being implemented to sink 4000 tube wells. The programme will be continued during 2005-2006 to provide 3500 tube wells in areas where there is groundwater potential.

Scheme for improving the efficiency of Agricultural Pump sets

The energy efficiency in farm sector pump sets is poor as the farmers are using higher capacity pump sets than the required level or using old pump sets. They have poor discharge compared to their horsepower. The improper installation of switchboards, earth and allied accessories also increases power consumption.

The efficiency of the electric pump sets can be increased by:

a. Replacing the old pump sets with new energy efficient ISI certified latest technology pump sets.
b. Providing proper electric accessories, mains, earths etc
c. Providing facility for ground water recharging by taking up Rain Water Harvesting measures around the well.
d. Decreasing the power consumption by reducing the pumping hours by providing Micro Irrigation facility, which allows drawl of less water from the wells.
e. Ensuring economy in using ground water by 30%.
f. Increasing the farm yield by the use of micro irrigation and fertigation cum herbigation so that the farmer will have higher income.

g. Revised cropping pattern by promoting integrated farming to increase the income of farmers. Reduced energy, water, pesticide and fertiliser consumption will also increase profitability of farmers.

It was therefore proposed to cover the dark and grey blocks of the state with this energy efficiency programme for electric farm pump sets based on the study conducted by Tamilnadu Agricultural University and Tamil Nadu Electricity Board.

During 2003-04, a pilot project have been taken up in three transformer areas in Theni, Coimbatore and Villupuram districts to study the effect on the consumption of energy and a total of 34 pumpsets have been energised at a cost of Rs.11.12 lakhs under this programme. During 2004-05, it is programmed to energise 20,000 pumpsets at a cost of Rs.10.50 crores. During 2005-2006 also, the programme will be continued.

SOIL AND WATER CONSERVATION PROGRAMMES

Soil and Water Conservation in the Catchments of River Valley Project

Sedimentation of reservoirs is a serious problem posing a threat to the life of the reservoir, its storage capacity, the command area and generation of power. Sedimentation is caused due to soil erosion and improper land use in the catchment areas. With a view to reduce siltation of the reservoirs, various soil conservation measures are taken up in the River Valley Catchments of Tamil Nadu.

The Soil Conservation works are taken up on watershed basis in the Very High and High priority watersheds identified by the All India Soil and land Use Survey Organisation (AISLUS) and approved by Government of India depending upon the rainfall, nature of erosion, soil texture, slope and topography of the area. During 2004-05, the programme is being implemented at a cost of Rs.734.41 lakhs covering 5965 ha. in South Pennaiyar & Mettur Catchments in Dharmapuri and Krishnagiri districts. During 2005 - 06, it is proposed to cover 6200 ha. in South Pennaiyar & Mettur Catchments at a cost of Rs. 1000.00 lakhs.

Soil Conservation in Tribal Areas

The objective of the scheme is to improve the economic status of the tribal farmers so as to lift them above the poverty line. By conserving the soil, improving the water holding capacity and improving the land use, the land productivity is proposed to be increased. Integrated development of tribal areas has been taken up under a multi sector approach in Jawadhu hills (Vellore and Tiruvannamalai district), Kalrayan hills (Salem and Villupuram districts) Shevaroy hills (Salem district), Sitheri hills (Dharmapuri district), Kolli hills (Namakkal district) and Pachamalai (Trichy district). Land Levelling, construction of contour rubble bunds with vegetative hedges and check dams are taken up in tribal areas at Government cost. These programmes provide employment to the local people and also help them to bring the treated areas under cultivation. During 2004-2005, this scheme is being implemented to treat 699 ha. of tribal lands at a cost of Rs.92.90 lakhs. During 2005-06, it is proposed, to implement the scheme to treat 600 ha. of tribal lands at a cost of Rs.92.90 lakhs.

Soil and Water Conservation under Hill Area Development Programme

Soil conservation scheme is being implemented in the Nilgiris District as an integrated watershed development programme, benefiting all the farmers irrespective of category and crops. The Nilgiris district has been delineated into 75 watersheds. Among these watersheds, highly degraded watersheds are taken up for treatment. Besides, soil conservation works, landslide preventive measures etc., have been taken up. Under this scheme, 100% grant is provided for soil conservation measures. However, beneficiaries are required to contribute at the rate of 10% for individual works and 5% for community works. In the case of SC/ST communities, the contribution required is 5%. The landslide treatment measures are however, executed under 100% grant. During the year 2004 - 05, this programme is being implemented to treat an area of 649 ha. with an outlay of Rs.358.01 lakhs. During 2005-06, it is proposed to treat an area 520.500 ha. with the financial outlay of Rs. 560.310 lakhs.

Soil and Water Conservation under Western Ghat Development Programme

Soil Conservation is one of the important aspects under Western Ghats Development Programme. The works are taken up in the Western Ghats Region covering the districts of Coimbatore, Erode, Dindigul, Theni, Virudhunagar, Tirunelveli, and Kanyakumari. During 2004-2005, Soil Conservation works in 645 ha. are being taken up at a cost of Rs.327.36 lakhs in eight districts. During 2005-2006, an area of 741 ha. is proposed to be covered at a cost of Rs.347.754 lakhs in eight districts.

Soil and Water Conservation Scheme for Rain Water Harvesting through Farm Ponds and Rejuvenation of Failed / Unused Wells

Over the years the area under well irrigation is increasing, resulting in increased ground water exploitation. The increased exploitation of ground water in many areas of the State has resulted in lowering of water table below the economic pumping level. Mining of ground water has occurred due to sinking of bore wells up to 1000 feet in certain areas. The ground water potential has been exploited to such an extent that special methods of rainwater harvesting and ground water recharge are warranted to save the well irrigated areas. Government have provided resilience to the drought affected agriculture by promoting rain water harvesting. They have sanctioned the construction of farm ponds and rejuvenation of failed / unused / abandoned wells. A total of 8833 farm ponds were constructed at a total cost of Rs. 2564.29 lakhs and 2093 wells have been rejuvenated at a cost of Rs.264.37 lakhs during 2003-04. An outlay of Rs.10.00 crores has been provided in the BE 2005-06 for taking up soil and conservation works. The Agricultural Engineering Department will plan rain water harvesting and run-off management on mini watershed basis. This proramme will be designed in such a way that the watershed will get the benefit of convergence of various watershed development and individual beneficiary oriented programmes. A dynamic monitoring and evaluation methodology will be inbuilt in the programme. Rain water harvesting will be given further importance and will be continued in 2005-06 under all the Soil and Water Conservation programmes and NABARD assisted Rain Water Harvesting under RIDF Programme. Rain Water Harvesting structures such as farm ponds, check dams and rejuvenation of abandoned wells are proposed to be taken up during 2005-06, with the NABARD assistance under Rural Infrastructure Development Funds so as to benefit 10,000 farmers.

Pilot Project for Development of Coastal Saline and Sand dune Affected Soils of Nagapattinam - Vedaranyam Zone

The Coastal areas of Nagapattinam district Tamil Nadu consists of sand dunes and sandy belt with fresh water overlying saline ground water. Sea water intrusion is noticed in the Vedaranyam canal. The Coastal Zone in between Nagapattinam and Vedaranyam has natural resources, but have problems posed by the sea water intrusion as well as the climatic conditions. This pilot project aims at developing the lands affected by sand dunes, sandy plains with coastal aquifer, the saline belt and the water logged areas in a more scientific way. The three-year project is being implemented by this department in association with the Tamil Nadu Agricultural University (TNAU) Coimbatore and with farmers using a Participatory Rural Appraisal (PRA) approach. Government of India has accorded sanction for Rs.1.20 crores which includes the Government share of Rs.89.92 lakhs and beneficiary share of Rs,30.08 lakhs for this three year project to cover an area of 2000 ha. During 2003-2004, the programme was implemented in 8 villages at an area of 792 ha. for Rs.31.18 lakhs. During 2004- 2005, this programme is being executed with the full .assistance of Central Government at a cost of Rs 34.217 lakhs in an area of 810 ha. During 2005-2006, the final phase of this programme will be implemented to cover the balance area of 398 ha. at a cost of Rs.24.73 lakhs.

Centrally Sponsored Scheme on Reclamation of Alkali Soils

The Centrally Sponsored Programme for Reclamation of Alkali Soils in the State aims at improving the productivity of the alkaline lands by adopting suitable package of alkali soil reclamation technology. An outlay of Rs.690.00 lakhs has been approved for reclamation of 5000 ha. of alkaline soil during the years 2003 - 2004 and 2004 - 2005. During 2003-2004, an area of 929.24 ha. have been treated at a cost of Rs.19.26 lakhs. During 2004-2005, this scheme is being implemented in 1527 ha. of alkaline soils with an outlay of Rs.50.00 lakhs. During 2005-2006, it is proposed to continue the programme to cover an area of 3054 ha. at a cost of Rs.100 lakhs.

REMOTE SENSING

Remote Sensing and Geographic Information Research Centre at the O/o. Chief Engineer, AED, Chennai

The Remote Sensing Centre of the Agricultural Engineering Department provides input for various Watershed Development Programmes and Wasteland Development Programmes. Under Western Ghats Development Programme, the status report for western ghats region are prepared annually on watershed basis depicting the Geology, Geomorphology, Infrastructure, Micro watershed codification and soil erosion for prioritisation of the watershed activities. This centre has been recognized by Anna University as Research Centre in which Ph.D, and M.S programmes of Anna University can be carried out by Department staff on external registration. This centre can also guide research programmes of Anna University. The State Land Use Board has given a study to prepare Watershed Atlases and Soil Watershed Atlases for all the districts in Tamil Nadu. Such maps have been used to delineate the micro water sheds under NWDPRA and Comprehensive Waste land development projects.

District GIS Centre

A Geographical Information System Centre has been created at Theni and Salem. This will help to build database of all the resources available at watershed level for taking up various development activities at micro watershed level using GIS Software. The Watershed Development Programmes like River Valley Project, IWDP, DPAP, NWDPRA, etc., are carried out by this department all over Tamil Nadu. It is proposed to have information on watersheds covered under various programmes on 1:10000 scales and create a GIS for each district. Maps will be created for each project area in which information upto the field level in the micro watersheds will be built including field boundary, SF No., land use, soil-cropping practices, etc. At the micro watershed level, runoff and soil erosion will be worked out and soil conservation works will be designed accordingly. Similarly, one GIS centre is proposed to be established in Trichy during 2005-06.

MICRO IRRIGATION

The Agricultural Engineering Department is engaged in popularising Micro Irrigation systems (Both Drip & Sprinkler) in Tamil Nadu, since the year 1985 - 86. Among the three sources of irrigation namely, canals, tanks and wells, the wells account for 51.70% of the net area irrigated. In order to reduce the stress on well irrigation and also to ensure judicious and economic use of irrigation water, through installation of Micro Irrigation Systems, viz. Drip / Sprinkler Irrigation Systems, an allocation of Rs 51.22 crores was made in 2004-2005 and the works are under progress. The scheme conceived by the State Government supports all the farmers for all the crops and will be continued. During 2005-2006, it is proposed to continue the programme with an outlay of Rs.17 crores. A plan has been laid for 'implementing Micro Irrigation system in State Seed Farms covering an area of 300 Ha during the year to popularise the Jetropha, Sweetswargham and Sugarbeet .

MINOR IRRIGATION

Tube Wells Scheme

Creation of additional irrigation facilities, stabilisation of the area already under irrigation and promotion of conjunctive use of surface water and ground water by harnessing groundwater resources are the objectives of the Tube Wells Scheme. The State of Tamil Nadu has nearly exhausted the surface water sources and there is increasingly more dependence on ground water. Therefore, sinking of new tubewells and revitalisation of dried up wells are taken up under this programme. Under this scheme, the Resistivity meters and Electrical loggers are provided to farmers for locating well sites and aquifers. Drilling Rigs, Mini Drills, Hand Boring Sets, Rock Blasting Units, Long Hole Equipment and Air Compressors are also provided to farmers on hire basis to take up these activities. The Department maintains a fleet of minor irrigation machinery of different types. During 2004-05, the programme is being implemented to sink 4000 tube wells. The programme will be continued during 2005-2006 to provide 3500 tube wells in areas where there is groundwater potential.

Scheme for improving the efficiency of Agricultural Pump Sets

The energy efficiency in farm sector pump sets is poor as the farmers are using higher capacity pump sets than the required level or using old pump sets. They have poor discharge compared to their horsepower. The improper installation of switchboards, earth and allied accessories also increases power consumption.

The efficiency of the electric pump sets can be increased by:

a. Replacing the old pump sets with new energy efficient ISI certified latest technology pump sets.

b. Providing proper electric accessories, mains, earths etc

c. Providing facility for ground water recharging by taking up Rain Water Harvesting measures around the well.

d. Decreasing the power consumption by reducing the pumping hours by providing Micro Irrigation facility, which allows drawl of less water from the wells.

e. Ensuring economy in using ground water by 30%.

f. Increasing the farm yield by the use of micro irrigation and fertigation cum herbigation so that the farmer will have higher income.

g. Revised cropping pattern by promoting integrated farming to increase the income of farmers. Reduced energy, water, pesticide and fertiliser consumption will also increase profitability of farmers.

It was therefore proposed to cover the dark and grey blocks of the state with this energy efficiency programme for electric farm pump sets based on the study conducted by Tamilnadu Agricultural University and Tamil Nadu Electricity Board. During 2003-04, a pilot project have been taken up in three transformer areas in Theni, Coimbatore and Villupuram districts to study the effect on the consumption of energy and a total of 34 pumpsets have been energised at a cost of Rs.11.12 lakhs under this programme. During 2004-05, it is programmed to energise 20,000 pumpsets at a cost of Rs.10.50 crores. During 2005-2006 also, the programme will be continued.

AGRICULTURAL MECHANISATION

Popularisation of Agricultural Machinery

AED provides support to farmers in reclamation of virgin lands and to carry out cultivation operations like ploughing, puddling and harvesting under the Agricultural Mechanisation Programme, assistance is provided to farmers for procuring farm equipments so as to to supplement the available "farm power", to reduce the drudgery of agricultural labourers, to ensure timeliness in carrying out various farm operations and to increase agricultural production. During 2004-2005, it is programmed to provide assistance to farmers for procuring 390 power tillers, 70 tractors, 5 Nos. of self propelled paddy transplanters, 5 nos. of self propelled reapers and 42 nos. of Novel agricultural implements like, Chisel Plough, Zero till seed drill, Rotavator, Self Propelled Reaper, Self Propelled Paddy Transplanter, Groundnut Harvester, etc. at a total subsidy amount of Rs.150.00 lakhs. This programme will be continued in 2005-06 also.

Provision of Deep Tillage and Earth Excavation Equipments

In this department 109 Nos. of D15A-15 Bulldozers and 125 Tractors are available for hiring out to the needy farmers at nominal hire charges. Besides Land Development, Agricultural Engineering Department implements several Watershed programmes including Wasteland Development and Drought Relief Programmes. These attachments are very much useful in opening hard pan and thereby increasing Water retention capacity and better aeration in soil and consequently facilitating increased crop growth. The demand from farmers for deep tillage, earth excavation equipments like backhoe are more and thereby the wasteland and soil eroded lands can be reclaimed. Government have provided Rs.44 lakhs and these equipments / machinery will be added to the fleet. These equipments will be used along with the Komatsu make excavators now used in Nagapattinam district, which were donated for Tsunami Relief Works for reclaiming Tsunami affected areas and waste lands in the State.

Agricultural Engineering Testing, Training & Demonstration Centres at 8 Workshop Premises of the Department

In view of the thrust given by the Government of India for popularisation of Agricultural Machinery and Micro Irrigation, it is proposed to provide testing and demonstration facility in the eight workshop premises of the department. Government have approved, the facility to be established in eight centres viz. Chennai, Madurai, Salem, Tiruvarur, Vellore, Coimbatore, Trichy and Tirunelveli for organising Demonstration and Testing of Agricultural machinery and Micro Irrigation system by the Agricultural Engineering Department, for the benefit of the farming community, at a cost of Rs.41.00 lakhs during 2004-05. It is also proposed to provide adequate infrastructural facilities like computers, demonstration and testing equipments etc. required for conducting of demonstration and testing of machinery / micro irrigation equipments in the eight centres during 2005-06 at a cost of Rs.20.00 lakhs.

Central Sector Scheme of Training Farmers in using Agricultural Machinery

In order to stream line the process of Farm Mechanization to increase Productivity, Production income of the farmers and at the same time to conserve the inputs for Crop production, it is considered necessary that the various levels of Man-power engaged in the area of Agricultural Mechanisation is imparted training and also exposed to refresher trainings for new technology components. With the above objective, the Government have approved a training programme for providing training to the farmers at nearby places of their residences. The Government have sanctioned a sum of Rs.3.50 lakhs for imparting training to the farmers during 2004-2005. It is proposed to conduct the training Programmes in Agricultural Tractor Workshops functioning under the control of Agricultural Engineering Department, in Coimbatore, Madurai, Trichy, Tirunelveli, Vellore, Thiruvarur and Chennai and also at Agricultural Engineering Training Centre at Trichirapalli during 2004-2005. Steps are being taken to train 200 farmers during 2004-05. During 2005-06, the training programme will be further expanded to more centres. The training programmes are taken up in close co-ordination with the Southern Region, Farm Machinery Training and Testing Institute, Garladinne, Ananthapur District, Andhra Pradesh to accelerate the pace of farm mechanisation in the State.

TAMILNADU WATERSHED DEVELOPMENT AGENCY AND COMPREHENSIVE WASTELAND PROGRAMME

Participatory Watershed Development

This scheme is under implementation in the wastelands (patta lands of farmers) since 2002-03 based on micro watershed approach with a provision of Rs.8,500/- per hectare as subsidy. The programme has been implemented as a pilot project in 10 Districts to cover 55,000 Hectares with a financial allocation of Rs.30 Crores in 2002-03. An area of 21,771 Hectares was covered under the project and 23,66,074 seedlings (15,39,267 Agro-Forestry and 8,26,807 Horticulture) were planted.

In addition to the 10 Pilot Districts, the programme was extended subsequently, to 13 Districts viz., Thoothukudi, Viruthunagar, Sivagangai, Madurai, Coimbatore, Erode, Karur, Perambalur, Villupuram, Cuddalore, Tiruvallur, Dharmapuri and Kancheepuram during 2003 - 2004 to cover 50,000 Hectares with a financial allocation of Rs.25 Crores.

During 2003-04 as on 31-03-04, an area of 23,600 Hectares has been covered in 10 Pilot Districts by planting 25,00,067 seedlings (6,87,013 Agro-Forestry and 18,13,054 Horticulture).

In the 13 new districts, the preliminary activities like registration of District Watershed Development Agency,

Selection of Non-Governmental Organisations, Registration of Village Development Associations conducting of Participatory Rural Appraisal Exercises, Identification of Entry Point activities, Training to Farmers, Identification of bore-well points etc., were taken up in 2003-2004. In 2004-05 this programme is being implemented in 60,000 ha of which 54,000 ha with Tree seedlings and 6,000 ha with Fodder Crops at an estimated cost of Rs.50 Crores in the following districts as detailed below :-

District-wise Planting Target for 2004-2005 (In Ha)

Sl. No.	Name of the District	Area under Tree Crops	Area under Fodder Crops	Total
1.	Tirunelveli	2100	400	2500
2.	Pudukkottai	2400	300	2700
3.	Dindigul	2100	300	2400
4.	Vellore	2400	200	2600
5.	Tiruvannamalai	2100	200	2300
6.	Salem	2400	400	2800
7.	Thiruchirappalli	3000	200	3200
8.	Theni	2400	20	2420
9.	Ramanathapuram	2400	10	2410
10.	Namakkal	1800	400	2200
11.	Thoothukudi	2100	200	2300
12.	Virudhunagar	3000	50	3050
13.	Sivaganga	2100	100	2200
14.	Madurai	2400	20	2420
15.	Coimbatore	2100	50	2150
16.	Erode	3000	210	3210
17.	Karur	3000	500	3500
18.	Perambalur	2100	600	2700
19.	Villupuram	2400	200	2600
20.	Cuddalore	2400	50	2450
21.	Tiruvallur	2100	350	2450
22.	Dharmapuri	2100	620	2720
23.	Kanchipuram	2100	620	2720
	Total	54,000	6,000	60,000

The P.Os DWDAs / JDAs have been permitted to raise up to 25% of the total requirement of Horticulture and Agro-Forestry seedlings from 2004-2005 onwards involving experienced Self Help Groups, Federation of Self Help Groups through the State Seed Farms, Oil Seed Farms, Coconut Nurseries and Pulses Farms run by the Department of Agriculture in their respective Districts itself with quality assurance, provided that the production cost does not exceed more than the Government approved production rate to be fixed by Commissioner of Horticulture and Plantation Crops (CH&PC) and Principal Chief Conservator of Forests (PCCF). This would facilitate the smooth implementation of the programme as detailed below:

1) Timely supply of seedlings to the Farmers on receipt of the seasonal rain fall.
2) Substantial saving in Transport charges.
3) Assessment of exact species and varieties required by the Farmers to raise and supply the seedlings accordingly.
4) To meet the seedling requirement for the casualty replacement (10%)
5) In the event of non-planting of seedlings raised due to failure of monsoon, to maintain the seedlings locally till next planting season.
6) To supply optimum aged seedlings as desired by the Farmers.
7) Further, advance planning to raise seedling so that the seedlings could be ready for planting when the rainy season starts.

The balance 75% of the Agro-Forestry and Horticulture seedlings to be procured directly by the POs / JDAs from the PCCF., CH&PC., TNAU, TAFCORN., and DRDA Nurseries.

The total requirement of Horticulture / Agro-Forestry seedlings for planting programme 2004-2005 under CWP have been assessed as 43.16 lakhs for 23 districts. This has been worked out based on the last year left over stock of seedlings and the 25% of the seedlings permitted to be raised locally in the districts through SHGs and Government farms as indicated above. In addition to the Tree crops, the Fodder crop and Jatropha, the bio-fuel crops have also been included in 2004-2005 planting programme under CWP.

In 2004-05, against the target of 60,000 ha. an area of 38,769 ha has been covered upto February 2005 of which 36,884 ha are under Tree crops and 1,985 ha are under Fodder crops. The subsidy norm for Tree crops is Rs.8,500/- per ha.

The subsidy norm for Fodder crop is Rs.8,400/- per ha for Annual fodder and Rs.7,500/- per ha for Perennial fodder. Subsidy norms for Jatropha is under the active consideration of the Government.

The Government have sanctioned a sum of Rs.72.49 Cr. for the CWP so far and a sum of Rs.49.24 Cr. has already been released to the 23 district Collectors / Chairman, DWDAs and HODs concerned so as to implement the CWP under planting programme through participatory approach.

In 2005-06 this programme will continue with a planting target of 60,000 ha. in 23 districts.

Leasing of Government Waste-lands to Corporate Houses

This scheme is also implemented in all the districts from 2002-2003 onwards except in Chennai, Kanyakumari and Nilgiris Districts. For leasing, the Government Wastelands have been classified into Blocks of lands less than 8 Hectares and the Blocks of lands 8 Hectares and above. Government Wastelands are being leased to Corporate Houses, Small Companies, Co-operative Societies, Federation of Self Help Groups and Self Help Groups for a period of 30 years.

The District level Committee under the Chairmanship of the District Collector, scrutinizes the applications and clears the lands for the category of less than 8 Hectares block. For 8 Hectares and above categories, the District Level Committee recommends the eligible firms to the High Level Committee constituted at State level under the Chairmanship of Chief Secretary to Government. The High Level Committee periodically meets and further scrutinizes the applications and their related project proposals of the firms and clear the lands.

The High Level Committee has recommended to the Government to lease Government Waste-lands to an extent of 1054.05.5 Hectares under 8 Hectare and above category of which G.O. has been issued for 624.77.5 Hectares. Under below 8 Hectare category, the District Collectors have so far cleared 291 Hectares. The Memorandum Of Understanding (i.e.,) the model lease deed copy has been communicated to all the District Collectors and signing the MoU with the firms concerned is in progress.

Drought Prone Areas Programme (DPAP)

The Drought Prone Areas Programme (DPAP) is one of the areas development programmes implemented in Tamil Nadu since 1972-73 to tackle the special problem faced by the fragile areas, which are constantly affected by severe drought conditions. The basic objective of the programme is to minimise the adverse effects of drought on the production of crops and livestock and the productivity of land, water and human resources thereby ultimately leading to drought proofing of the affected areas. Since 1995-96, this programme is being implemented as per the guidelines of Government of India on Watershed basis with peoples' participation. Each Watershed covers approximately an area of 500 ha.

At present, this programme is being implemented in 80 blocks of 17 districts in Tamil Nadu. The Government of India and the Government of Tamil Nadu shared the expenditure on 50:50 basis for the projects sanctioned upto 31-3-1999. The funding pattern has been revised as 75:25 between Central and State Governments for the projects sanctioned from 1.4.1999. The unit cost for developing a watershed has been fixed at Rs.6000/- per ha. with effect from 1.4.2000 and the project period is 5 years.

Under Drought Prone Areas Programme watershed development projects are sanctioned by Government of India in batches. In the first batch, 297 Watershed projects were sanctioned and completed during the period 1995-2000 at a total cost of Rs.5920.25 lakhs. (No project has been sanctioned in II, III and VI batches for Tamil Nadu). The total area treated is 1,48,911 Ha. Presently IV, V, VII, VIII and IX Batches of DPAP watershed programmes are under implementation. During 2004-05, the Government of India have sanctioned the X Batch DPAP Watershed Programmes for implementation in 160 watersheds spread over in 17 districts at a total cost of Rs. 4800 lakhs for the period 2004-05 to 2008-09.

Batch-wise number of watersheds sanctioned, total allocation of funds, funds released by Central and State Governments and expenditure details are given below:

NATIONAL WATERSHED DEVELOPMENT PROJECT FOR RAINFED AREAS (NWDPRA)

Restructured NWDPRA for Xth plan is a Five Year project starting from 2002-03 to 2006-07. It is one of the components of Macro management mode. The scheme is implemented in blocks having less than 30% assured means of irrigation in arable lands and having slopes less than 8%.

Objectives: The broad objectives of the NWDPRA are as follows:

- Conservation, development and sustainable management of natural resources including their use.
- Enhancement of Agricultural productivity and production in a sustainable manner.
- Restoration of ecological balance in the degraded and fragile rain fed eco- systems by greening these areas through appropriate mix of trees, shrubs and grasses.
- Reduction in regional disparity between irrigated and rainfed areas.
- Creation of sustained employment opportunities for the rural community including the landless.

NWDPRA is implemented in 23 districts of Tamil Nadu except Thanjavur, Nagapattinam, Tiruvarur, Chennai, Nilgiris and Kanyakumari. The unit cost per ha. is Rs.4500/-. As per Government of India guidelines the district watershed committee headed by the District Collector, as chairman have been formed, district nodal agency have been positioned. Selection of PIAs and Watershed Development Team members completed. For 763 micro watersheds, 102 Nos. of Project Implementation Agency (PIA) have been identified and engaged. Departments like Agriculture, Agricultural Engineering, Forest, and NGOs were selected as PIAs in all 23 districts. The delineation of micro watershed have been completed with the help of Agricultural Engineering Department.

During first year of Xth plan period 2002-2003, the amount released by the State Government for 763 micro watershed for the year 2002-03 has been fully utilized. During 2003-04, an amount of Rs.1525.00 lakhs is sanctioned and released for the treatment of 33.888 ha. and the entire amount has been fully utilized. During 2004-05,as per G.O.M.S.No 230 Agriculture (WLD) Department dated: 14.7.2003 NWDPRA has been taken in to the fold of TAWDEV Agency both at state and district level with effect from 30.9.04.

This year 2004-05 vide G.O.Ms.No.201 Agri. (B&P) Dept Dated:11.6.2004 an amount of Rs.2000.000 lakhs has been sanctioned to treat an area of 44444 Ha.

WATERSHED DEVELOPMENT FUND

Watershed Development Fund has been created in NABARD. The object of the fund is to spread the message of participating watershed development. The fund will be utilized to create necessary framework conditions to replicate and consolidate the isolated successful initiatives under different programmes in the Government, Semi Government & NGO sectors. The WDF is operationalised in close condition with Central & State government. There are 2 components in WDF. 1. Grant component (1/3), and Loan component (2/3). Repayment period is 9 years and it commences 3 years after availing the loan. The interest rate is 4.5%. Cost norm is Rs.6,000/- per ha. At present, WDF being implemented in 7 districts namely, Pudukottai (5), Vellore (2), Ramnad (3), Sivaganga (3), Tuticorin (4), Dindigul (7), and Tirunelveli (3) in coordination with NABARD. The WDF is monitored and guided by State Level Steering Committee under the Chairmanship of Agriculture Production Commissioner.

10

AGRICULTURAL EDUCATION AND RESEARCH

INTRODUCTION

The Tamil Nadu Agricultural University (TNAU) has been constantly orienting its activities of agricultural education, research and extension towards the goal of not only maintaining the tempo of self sufficiency in food production but exploring the agricultural potentials also for export of agricultural produce from our state and thus improve the rural economy of Tamil Nadu.

The agricultural scenario in Tamil Nadu has changed fast with new problems and challenges. To be specific, the yield levels of major crops have already reached a plateau, water resources have become more scarce, the gross cropped and gross irrigated area are marginally decelerating while the era of scarcity has given rise to the era of surplus which has to be marketed successfully both internally and externally. To meet these challenges TNAU has reoriented its policies as well as programmes and strategies.

EDUCATION

Globalisation has created new opportunities and problems to Indian farming sector. Indian farming sector needs competent technical manpower to meet their challenges of liberalization. Agricultural Education in general and its post graduate education in particular has to produce technical manpower to meet the changing needs of the stakeholders. Hence, more emphasis is given for continuous upgradation of syllabus, introduction of need based courses taking up collaborative research with public and private sector and introduction of new teaching methods.

Undergraduate programmes in Agriculture, Horticulture, Forestry, Home Science and Agricultural Engineering are offered in the ten constituent colleges of the University located in seven campuses (Annexure I). Apart from the regular degree programmes of B.Sc.(Ag.), B.Sc.(Hort.), B.Tech (Agrl.Engg.), B.Sc.(H.Sc.), B.Sc.(Forestry), three self supporting Bachelor of Technology programmes in Food Processing Engineering, Horticulture and Agrl. Biotechnology are also being offered since 2001. During 2004 one more new self supporting technology course namely B.Tech. (Energy and Environmental Engineering) and three post graduate programmes namely M.Sc. (Bio informatics), M.Sc. (Bio Chemical Technology) and M.Sc. (Agricultural Meteorology) have been introduced.

The objective of providing technically skilled manpower to the task of modernising the agriculture in the state is achieved by offering the students the following courses and training programmes as part of the curriculum.

1. Every course offered to the undergraduate students has been modified so as to include practical component in order to develop the students' ability in applying the knowledge to solve the field problems.
2. Commercial Agriculture courses like mushroom cultivation, nursery preparation, poultry production, production of biopesticides, biofertilizers, bee keeping etc are offered so that the students could be involved in the whole process of production. This hands on training will be helpful for developing self confidence and an urge among the students to become entrepreneurs .
3. The students are provided opportunities to interact with the progressive farmers, government development department personnel and NGOs in order to enable them to acquire themselves with the problems faced by the stakeholders in the field of agriculture.

4. The students are exposed to real farming situations by making them stay with the farmers for 105 days under Rural Agricultural Work Experience programme.
5. The students are given coaching in specialised subjects and general knowledge including English to help them succeed in various competitive examinations, Special coaching classes are conducted for GRE, TOEFL by the Directorate of Students Welfare to help them in getting admission for higher degree programmes in foreign universities.

Master's Degree Programmes in 22 disciplines and doctoral degree programmes in 20 disciplines (agriculture and allied) are being offered in the University. In the Post Graduate programme while fixing the topics to take up research, the field problems faced by the farmers are taken into account and the entire research is oriented to solve such problems.

During 2004-2005, 293 boys (51%) and 277 girls (49%) were admitted in the Under Graduate Programmes. In the Post Graduate Programme, 114 boys (46%) and 132 girls (54%) were also admitted. While in the Ph.D Programme 36 boys (49%) and 38 girls (51%) were admitted. Out of the 890 students in various programmes, 443 were boys (50%) and 447 were girls (50%).

At Dr.C.Subramaniam, Institute of Agriculture situated in the campus of Oil Seeds Research Station, Tindivanam one Diploma course in Agriculture with an intake of 50 students / year has been started during 2004. It is a self supporting course.

RESEARCH

A total of 507 research schemes are being undertaken by the scientists of TNAU with the funds provided by the State and Central Governments, Foreign and Private agencies.

The research programmes are spread over the seven teaching campuses, 32 research stations and 13 transfer of technology centers located in all the seven agro-climatic zones of Tamil Nadu, in order to meet the location specific needs of the farming community (Annexure II).

Future Plans

Research work is being focused on the following lines.

- Introduction of bio-energy crops viz., sugarbeet, sweet sorghum and Jatropha, standardization of agrotechniques for growing jatropha and efficient extraction of oil from such crops, intensification of research for sugar extraction from sugar beet and alchol production from sweet sorghum and dissemination of such technologies to farmers. Pilot plant for bio-diesel extraction has been established for demonstration purpose.
- Development of sustainable organic farming and precision farming systems.
- Evolving an integrated farming system for sustainable agriculture through farmers participatory research.
- Rainfall prediction for south west monsoon and north east monsoon seasons and weather based farm advisories through medium range weather forecast.
- Selection of crop genotypes for abiotic stress tolerance (for heat, salinity and drought).
- Evolving systems for recycling and to study the impact on industrial and urban wastes for crop cultivation including selection of a suitable crop for such affected soil.
- Prescribing fertilizers based on soil test crop response for specific yield target.
- Utilizing remote sensing and GIS data base for farm planning in Tamil Nadu.
- Advancing high yielding rice lines developed through biotechnology.
- To develop rice hybrids and varieties with fine grain quality and rice varieties with drought tolerance through biotechnology.
- Evolution of transgenic rice lines resistance to stemborer/ leaf folder.
- Development of rice varieties of short duration and quality, medium duration with higher yield and superior

quality, CMS based non aromatic hybrids with good grain quality and varieties with high degree of resistance.

- Evolution of hybrids based on CMS lines in pearl millet and sorghum, single cross hybrids in maize and varieties in ragi and millets suitable for arid tracts.
- Evolution of genetic male sterility based hybrids, early duration redgram varieties with resistance to pod borer, yellow mosaic virus resistant varieties in black gram and green gram and development of short duration soybean varieties suited for inter cropping, suitable lab lab and cowpea varieties for vegetable purpose.
- Development of groundnut varieties suitable for rainfed / irrigated conditions with high yield and oil content, sesame varieties with drought resistance and oil content, sunflower hybrids / populations with high yield and oil content, castor varieties / hybrids of suitable for pure and intercropping systems.
- Development of cotton varieties / hybrids suitable for rainfed / irrigated / rice fallow conditions in upland cotton and desi cotton varieties with long lint and high fibre strength, hybrids with resistance to boll worm.
- Development of high yielding and nutritive cereal / legume fodder.
- In major Agricultural and Horticultural crops including vegetables, the main focus would be on development of pest and disease resistant varieties and hybrids. New hybrids in tomato which are resistant to major disease which can be grown successfully without spraying pesticides have been developed and the same will be released for cultivation by farmers during 2006. These hybrids have a potential yield of 90-100 t/ha without pesticide application.
- Development of fruit borer resistant hybrids in brinjal through conventional breeding and varieties of local preference having Bt gene with high level of expression of resistance to shoot and fruit borer.
- To develop hybrids in banana with nematode and disease tolerance.

- To evolve integrated management practices to control soil borne pathogens in banana.
- Identification and development of new strains of biocontrol agents.
- Developing pest management strategies in transgenic (Bt) cotton.
- To identify and multiply suitable fruit, timber and non timber species for growing in waste lands since wasteland constitutes major area of arable land.
- Medicinal plants are being given special attention. In addition to standardization of cultural techniques for important medicinal plants such as Coleus forskholli, Gloriosa, Phyllanthus, Withania and Andrographis, identification and extraction of alkaloids and secondary metabolites are also being undertaken.
- Alternative cropping strategies for extreme conditions of severe drought, moderate drought and normal monsoon in all the seven agro climatic zones of Tamil Nadu State have been identified and recommended.
- Post harvest technology centre has been set up aiming at development of entrepreneurs in Agriculture. Necessary infrastructural facilities were developed for use by the progressive farmers to avail the facility of product manufacturing from their agricultural produce especially during the time of glut.
- Development of microbial consortia for maximizing the crop productivity.
- Microbial technology for value addition of agricultural produces.
- Identifying newer formulation of inoculants for more shelf life.
- Isolation of acid tolerant nitrogen fixing cyanobacterial cultures for paddy
- Under Part II Plan, 11 research programmes namely maximising rice yield in rainfed areas of coastal district, establishing systems for outdoor cultivation of mushroom, demonstration of pesticide free cultivation

of tomato with virus disease resistant hybrids developed by TNAU, development of agro technologies for bio fuel crop, jatropha curcus, non-chemical methods of pest control in cole crops, eco friendly management of waste water, documentation of 'Low cost' and 'No cost' agro technologies of dry tracts, change in cropping pattern for export oriented agriculture etc. have been taken up during 2004-2005.

- With a view to supply quality seeds to the farming community and to support the state development departments connected with Agriculture and the Seed Industry, Tamil Nadu Agricultural University is supplying breeder seed for further multiplication and distribution. A total quantity of 1,66,779 kg of breeder seeds has been supplied. The details of breeder seed produced and distributed in the year 2003-04.

A total number of 16 new research schemes have been proposed to State Planning Commission under Part II plan of Government of Tamil Nadu (2005-2007) to emphasise the following lines of work.

1. Integrated soil nutrient management through vermicomposting
2. Management of the sugarcane woolly aphid in Tamil Nadu,
3. Introduction and popularizing tropical sugar beet in Tamil Nadu
4. Popularizing sweet sorghum in different agro climatic zones of Tamil Nadu through farmer's participatory research and demonstration
5. Strategies and programmes for increasing the production and productivity of maize in Tamil Nadu through farmer's participatory research and demonstration
6. Introduction of mini mobile sprinklers for rainfed groundnut cultivation by utilizing the shallow springs in the coastal belts of Ramanathapuram districts
7. Development and performance evaluation of ecofriendly bioformulations for the management of Coleus (Coleus forskohlii) root rot disease

8. Genetic enhancement of nucleus seed of rice varieties
9. Development of mycorrhizal biofertilizer and quality control unit for agricultural and horticultural crops
10. Eco-friendly recycling of primary treated distillery effluent in Jatropha curcas cultivation for greening the wasteland
11. Standardisation of organic production technology for turmeric
12. Physiological approaches to improve the productivity of pulses
13. Strengthening of quality seed production of millets, pulses and cotton
14. Introduction and popularisation of forage crops and technologies suitable for different agro climatic zones of Tamil Nadu
15. Improving the processing of Turmeric without using chemicals
16. Taking into account the severe damage caused by the recent tsunami to soils in the coastal districts due to accumulation of salt it has been proposed to multiply seeds of salt tolerant rice varieties developed by TNAU at Trichy and Aduthurai and distribute to farmers of coastal districts through Department of Agriculture.

POLICY INTERFACING

TNAU keeps on assessing the crop situations, rainfall and pace with the pattern of agricultural growth in the state. Based on these assessments, several agricultural development policy proposals have been developed and presented to the Department of Agriculture. These are incorporated in policy formulations after discussion with the extension officials.

EXTENSION

The Directorate of Extension Education of Tamil Nadu Agricultural University is primarily responsible for expeditious transfer of the latest technologies emanating from various research programmes of TNAU to the farming community and the Extension personnel of the State Agrl. Department, through its Krishi Vigyan Kendras, Plant Clinic Centres, Communication Centre and Training Division (Annexure III). During 2004 seven KVKs sponsored by ICAR have been added in 7 districts of Tamil Nadu. The communication centre at Coimbatore disseminates the innovations through mass media like radio, dailies, farm magazines, publications, video programmes and Doordarshan. It also conducts distant learning programmes like Farm School on AIR for the benefit of the farming community. A holistic approach for demonstrating precision farming is being implemented in Dharmapuri and Krishnagiri districts for getting maximum output for unit input spent. Preliminary results revealed that farmers are much convinced on the technologies demonstrated in this programme.

The focus of activities of the Directorate of Extension Education will be:

- Organizing Hi-Tech training to extension functionaries of Government, Quasi Government, NGOs, Bank officials etc.
- Imparting training to practicing farmers farm women, rural youths and FDG convenors.
- Organize pre monthly and monthly zonal workshops for the extension scientists of Department of Agriculture and Horticulture.
- Arrangement for farmers visit from state and neighbouring states, conduct State Level Farmers Day and Farmers Scientists interaction meet.
- Strengthening and monitoring of existing KVKs and PCCs.
- Publication of News letter and Valarum Velanmai.
- Other related need based Extension activities.

Distance Education

Though the Tamil Nadu Agricultural University was established to impart formal education, it was found that if the technology is to be transferred fast it cannot be done through formal education alone. The need has arisen that it has to be transformed into a dual and convergent mode of University.

The major goals of programme is to transform the Tamil Nadu Agricultural University from its formal education system to a dual and convergent mode system by adopting appropriate policies and systems of Open Distance Learning and Information and Communication Technology. In this direction TNAU is seeking assistance from Common Wealth of Learning (COL) (i) by way of getting software (ii) by way of learning materials and training in new areas such as :

- Integrated farming systems
- Rain water harvest and dryland improvement
- Value addition in farm produce at village level
- Post harvest technology and storage aspects
- Improving leadership qualities of farm women and rural sewage farms

SEED CERTIFICATION

Seed contains in itself the blue print for the agrarian prosperity in incipient form. It is the basic input for increasing agricultural productivity. The seed industry in the state is developing at a quick pace with the added importance provided by the Government year after year, for increasing the production of quality seeds. For protecting the farming community, a strong seed certification programme and a quality control mechanism are a must. Making available adequate quantity of quality seeds at the doorsteps of the farmer at the right time is essential for increasing agricultural production. In order to achieve this, this department is implementing Seed certification, Seed quality control, Seed testing and Training schemes for technical personnel, producers and dealers.

SEED CERTIFICATION

Seed certification is done to encourage the production of adequate quantities of genetically pure and good quality seeds, particularly of high yielding improved varieties and hybrids of different crops. In order to achieve this, the Seed Certification Department carries out the certification work in various seeds notified under Indian Seeds Act, 1966 and in accordance with the provisions of the Seed Rules 1968 to maintain the quality of the seeds produced in the state. In early eighties 18,000 metric tonnes of Certified seeds were produced. During the end of the Eighth Plan Period, it was further enhanced to 24,452 metric tones. Later a remarkable increase was recorded by producing 41,331 metric tonnes in the year 2003-04. About 15 percent of the total seed requirement is covered with certified seeds. The crop wise certified seed coverage shows 37 percent with Paddy, 7 percent with Millets, 48 percent with Cotton, 12 percent with Pulses and 7 percent with Vegetables. In addition, about ten percent of the seed requirement is met with truthful labeled seeds.

The past experience has shown that an increase in the production of certified seeds by private sector was very much useful in increasing the total certified seed production. Hence, the private sector seed production needs to be promoted further, in order to fill up the gap between availability and requirement of quality seeds. During 2004-05 as against the target of 46,500 M.T. a quantity of 37,898 M.T. seeds were certified up to February '05. It is proposed to increase it to 48,500 M.T. during 2005-06.

SEED QUALITY CONTROL

With ever increasing demand for seed, the basic input of agriculture, seed trade has to be constantly monitored by inspecting the seed dealer points periodically, and having a constant vigil on the marketing of seeds by taking seed samples in the seed lots distributed to the farming community. This is more valid in the current scenario where the private sector is playing a considerable role in the production and distribution of seeds.

The seed samples drawn are analysed in the notified Seed Testing Laboratories. Based on the results legal action or departmental action is being initiated against the defaulters. In accordance with the provisions of Seeds Act 1966, Seed Rules 1968 and Seeds (Control) Order 1983, the seed quality control wing of this department is regulating the quality of seeds sold. Besides this, the seed quality control wing is issuing the licenses for Seed dealers under Seed (Control) Order 1983.

The number of inspection, which was below 10,000 per year till 1990-91, had increased in early nineties and it reached 26,322 inspections at the end of the Eighth Plan Period. Action was initiated to maintain the status quo during the Ninth Plan Period. The annual average number of inspections per selling point had gone up from two to five in spite of increase in number of retail outlets. As a result 29,382 seed selling point inspections were made and 21,685 seed samples were drawn for quality cheek during 2003-04. During 2004-05 as against the target of 28,800 seed selling point inspections 28,639 inspections were made up to February '05 and as against the target of 21,500 seed samples to be drawn 21,536 seed samples were drawn up to February '05. It is proposed to make 30,000 seed selling point inspections and to draw 22,000 seed samples for quality cheek during 2005-06

SEED TESTING

Seed testing plays a pivotal role in modern agriculture. It is being carried out to analyse the factors like germination, physical purity, moisture, seed health and admixture of other distinguishable varieties. Seed testing is carried out in the notified seed testing laboratories. The Seed testing results are very important for the successful implementation of seed certification programs and seed law enforcement programs. Certified seed samples, Official seed samples from quality control wing and the Service samples sent by the farmers, seed dealers and seed producers are tested in the laboratories.

During 2003-04, 47,991 numbers of seed samples were tested. During 2004-05 as against the target of 60,000 seed samples, 43,669 seed samples were tested up to February '05 in the laboratories. During 2005-06 it is proposed to test 65,000 seed samples. To determine the genetic purity of a given seed lot, the grow out test is conducted at Kannampalayam (Coimbatore) farm and at the glass house attached to the Directorate of Seed certification. This test is conducted for the certification of parental, hybrid and foundation class cotton seeds, castor hybrids and also for the samples received from the Seed Inspection wing.

TRAINING

To promote quality seed production and distribution the following training programmes are organised by the Training Wing of this department.

Orientation Training

Training is given to the new technical officers of this department on seed certification seed testing and seed quality control.

Refresher Training

The already positioned technical officers of this department are trained on the latest techniques on seed production and on identification of newly released varieties.

Training to Seed Producers

The training is given on the seed production to seed producers. The training includes seed growers who are mostly small and marginal farmers.

Quality control Training to Seed Dealers

Training is given to the seed dealers on quality maintenance in storage, selling of seeds and on the regulatory aspects of seed legislation.

The number of beneficiaries under various training programmes was 28,680 during the year 2003-04 and during 2004-05 the number of beneficiaries was 30,308 upto February'05. During 2005-06 it is proposed to train 28,000 persons.

Achievements During 2004-05

During 2004-2005 a Provision of Rs. 6.78 lakhs have been made to create facilities for testing genetically modified crops for protecting the interests of the farming community. A sum of Rs.2.25 lakhs have been allocated to strengthen the training wing of the Department of Seed Certification for the purpose of providing digital camera and LCD Projector for imparting training to farmers, seed growers, seed producers, seed dealers and technical staff of this department. A sum of Rs.0.07 lakh has been utilised to provide telephone facilities to the training centre. To upgrade the existing facilities in seed testing a sum of Rs.1.60 lakhs is being utilised to the Seed Testing Laboratory attached to Directorate of Seed Certification to provide Generator. A sum of Rs.0.60 lakhs is utilised to strengthen the power main of the directorate.

Plan for 2005-06

During 2005-06 it is proposed to strengthen the facilities of genetic purity testing farm located at Kannampalayam at a cost of Rs.5.75 lakhs by construction of ground level tank (Rs.4.00 lakhs) and by provision of borewell (Rs.1.75 lakhs) for the purpose of protecting the interests of the farming community by way of carrying out grow-out tests throughout the year. To hasten the process of seed certification and to improve the efficiency of the department, the seed analytical results are declared through internet. A sum of Rs. 4.45 lakhs is proposed for upgrading the existing computer facilities in the Seed Testing Laboratories situated at Coimbatore, Dharmapuri, Trichy, Tanjore, Madurai, Tirunelveli and Kancheepuram, Deputy Directors of Seed Certification, Coimbatore and Madurai and for providing fax machine to Deputy Director of Seed Certification, Coimbatore. To ensure quality seed distribution to the farming community by way of curtailing spurious and substandard seed distribution it is proposed to install computers in five Assistant Directors of Seed Inspection offices located at Coimbatore, Salem, Madurai, Tirunelveli and Trichy at a cost of Rs.3.50 lakhs. It is proposed to improve the infrastructural facilities in the office of Director of Seed Certification at a cost of Rs.5.34 lakhs by way of rectification of approach road (Rs.2.00lakhs) construction of two-wheeler shed (Rs. 2.85) and providing basic amenities in training hostel (Rs.0.49 lakhs). An exclusive telephone lease line (under MLLN Date Circuit) is proposed to be installed at a cost of Rs.0.19 lakhs for maintaining the website of Department of seed Certification.

11

Agricultural Marketing and Agri-business

INTRODUCTION

Marketing infrastructure plays a pivotal role in fostering and sustaining the tempo of rural economic development. Marketing is as critical to better performance in Agriculture as farming itself. In order to create the needed infrastructure, promote institutions and also to regulate the marketing practices, Government of Tamil Nadu enacted the ' Tamil Nadu Agricultural Produce Marketing (Regulation) Act 1987 and Rules 1991 ' replacing the earlier Act 1959. The recent Act envisaged the organization of market committees in all the districts for the purpose of notifying the commodities, establishing regulated markets and assembling centres. The Department of Agricultural Marketing, which is functioning since 1977, with the main objective of Regulation of Agricultural Marketing, was renamed in the year 2001 as Department of Agricultural Marketing and Agri. Business in order to focus on other activities like Agri Export, Post Harvest Management, Food Processing, etc. The main activities of the Department of Agricultural Marketing and Agri. Business are as follows:

1. Establishment and maintenance of regulated markets in order to facilitate buying and selling of agricultural produce for the benefit of the farming community.
2. To create awareness among the farmers about the benefits of marketing their produce through regulated markets by publicity and propaganda.

3. Commercial grading of agricultural produce in the regulated markets and at farm holdings to help the producers to get remunerative price for their produce.
4. To take up Agmark grading of agricultural, animal husbandry and forestry products for the benefit of the consumers.
5. To set up Agriculture Export Zones for promoting export of agricultural produce by increasing the area under exportable crops, providing necessary post harvest management and other infrastructure required and information on prices prevailing at international markets as an integrated approach.
6. To set up modern cold storage facilities to enable the farmers to store and sell their produce at favourable price level (Cold chain from farm to market).

TO PROMOTE FOOD PROCESSING INDUSTRIES.

Market Committees

At present 20 Market committees are functioning in Tamil Nadu covering all districts except Chennai and the Nilgiris. Now, these Market Committees are functioning with nominated Members and Chairman / Chairperson elected by the members from January 2003.

Regulated Markets

There are 273 Regulated Markets, 15 Sub Markets, 15 Check Posts, 108 Rural Godowns and 108 Grading Centres functioning under the 20 Market Committees.

Competitive and remunerative prices are ensured for the produce sold by the farmers through closed tender system in the Regulated Markets. Free Grading facilities for agricultural commodities are also made available in the Regulated Markets.

The Market Committees collect 1% of the value of the produce transacted as market fee from the traders. Licence fee is also collected from the traders and weighmen. No fee is collected from producers / sellers for any service rendered in the Regulated Markets. The details of Market Committee-wise Regulated Markets and Grading Centers

Services Rendered in the Regulated Markets

Correct weighment, godown facilities, banking facilities, immediate payment, price information, rest sheds, drinking water facility, cattle sheds, free medical aid to farmers, input shops, payment counter for easy disbursement, phone and fax facilities, etc. are the services rendered in the regulated markets.

The Market Committee-wise infrastructure facilities provided in the Regulated Markets are shown in Table- 2.

Notification of Produce

So far, 40 agricultural commodities like cereals, pulses, oilseeds, cotton, turmeric, etc. have been notified.

Commercial Grading

To help the producers to get better prices for their produce according to their grades, 96 Commercial Grading Centres, 11 Kapas Grading Centers and one Tobacco Grading Centre are functioning in the Regulated Markets. Commercial grading of agricultural produce is done in the villages also. These centers are now renamed as "Post Harvest Technology Information Centers". The farmers are educated in adopting post harvest technologies on agricultural commodities by the technically qualified staff working in the above centers.

Pledge Loan Facilities to Farmers

In order to avoid distress sales by the small and marginal farmers in the peak season, Regulated Markets are issuing pledge loan to farmers. Under this scheme, the farmers can store their agricultural produce in the godowns of Regulated Markets for a maximum period of 6 months and take pledge loan of 75% of the total value of the produce upto a maximum of Rs.50,000.

Pledge Loan Facilities to Traders

As an incentive to the traders and to help them to make immediate payment to the farmers, pledge loan is given to the licensed traders in the Regulated Markets. Pledge loan of 50% of the total value of the purchased produce or Rs. 1.00 lakh, which ever is less, is made available to the traders. The maximum stocking period is 3 months for the traders.

Tamil Nadu Farmers Development and Welfare fund Scheme

Under this scheme, the farmers / tenants who sell their agricultural produce of one metric tonne or more through Regulated Markets every year are eligible for a grant of a lumpsum amount upto Rs.75, 000, in case of death/ permanent disability occurring due to accident. In the event of death the nominee of the farmer / tenant and in the absence of nominee, the legal heir of the family will be eligible for the lumpsum grant. The farmers need not pay any premium for this fund. The Market Committee concerned and the Tamil Nadu State Agricultural Marketing Board will bear the premium amount of Rs.10 per individual per year equally.

Agmark Grading

Grading under AGMARK has already become a symbol of quality. In order to help the consumer to get quality food products, 30 State Agmark Grading Laboratories, 15 Agricultural Officer (Marketing) Centers and one Principal Laboratory are functioning in the State. Agmark grading is done for centralised and de-centralised commodities by the technically qualified staff. Agmark labels are issued to the authorised packers under the direct supervision of the staff for certifying the quality and purity of the food products.

Construction of Drying Yards in the Villages

In order to minimise the post harvest losses in grains, the department has taken up construction of drying yards at village level. Under this scheme, so far 730 drying yards have been constructed at a total cost of Rs.13.24 crores. Construction of 200 drying yards at a total cost of Rs.4.00 crores for the year 2004-05 is in progress.

CONSTRUCTION OF MARKET COMPLEX

Paddy Market Complex

At Madurai, a market complex is being established exclusively for paddy at a total cost of Rs.12.60 crores. This market complex will cater to the marketing needs of paddy growing farmers and traders.

Turmeric Market Complex

Government orders were issued to establish a Turmeric market complex exclusively for Turmeric in Nassiyanur and Villarsampatti villages near Erode at a total cost of Rs.36.32 Crores. The proposed market complex will cater to the marketing needs of the farmers of the State and traders of the entire country. So far an extent of 49 acres in the villages of Nassiyanur and Villarsampatti have been purchased by the Turmeric Traders for establishing Turmeric Market Complex.

AGRICULTURAL PRODUCTION AND MARKETING INFORMATION CENTRE

Market intelligence plays vital role in marketing of Agricultural produce. If information on commodity prices, market trend, demands etc. of various markets is made available, the farmers could decide judiciously where and when to sell their agricultural produce to get maximum return. Taking this objective into consideration, establishment of Agricultural Production and Marketing Information Centre is in progress in 14 regulated markets at a cost of Rs.45.00 lakhs. Based on the success and usefulness of the facility, this scheme will be further expanded.

AGRI EXPORT ZONES

Agri Export Zones Established

To promote Agricultural Exports from Tamil Nadu, three Agri Export Zones for specific commodities have been established as detailed below.

Export Zone for Cut Flowers in Dharmapuri District

An Agri Export Zone has been established exclusively for cut flowers in Dharmapuri and Krishnagiri districts at a project cost of Rs.24.85 crores. TANFLORA, a Joint venture company of TIDCO provides infrastructure facilities like Central Packing House comprising of sorting, grading, packaging and marketing facilities, land on lease to farmers for setting up floriculture units at Hosur. This Agri Export Zone started exporting cut flowers to various countries from the year 2003-04 to the tune of 400 MT worth Rs. 10.80 Crores.

Export Zone for Flowers in Nilgiris District

An Agri Export Zone for flowers in The Nilgiris has been established with the participation by private sector at a project cost of Rs. 15.89 crores. Memorandum of understanding (MoU) for this Agri Export Zone has been signed between APEDA and Government of Tamil Nadu Agriculture Department on 6.2.2003. Infrastructure facilities like cold storage, refrigerated vans, common marketing facility, etc., would be created shortly. The Department of Horticulture & Plantation Crops is carrying out floriculture training to growers and providing cuttings at subsidized rate, subsidy to Green house, shade nets to growers. The Department also established information centre. The zone commenced its export from the year 2003-04.

Export Zone for Mango in Theni and 5 other Districts

An Agri Export Zone exclusively for mango has been established in Theni and five other districts namely Madurai, Virudhunagar, Tirunelveli, Dindigul and Kanniyakumari with private sector participation at a project cost of Rs.24.60 crores. Memorandum of understanding (MoU) for this Agri Export Zone has been signed between APEDA and Government of Tamil Nadu Agriculture Department on 6.2.2003. M/S. Maagrita Export Limited affirm registered by a private entrepreneur is to provide facilities like collection centres, processing units, cold storage, etc., shortly. Export from this zone will be commenced from middle of the year 2005.

Export Zone for Cashew in Cuddalore District

It is proposed to establish an Agri Export Zone exclusively for cashew at Cuddalore district through private sector participation at a project cost of Rs. 10.36 crores. This proposal has been approved by Government of India.

AGRI EXPORT ZONES PROPOSED

Export Zone for Banana in Tiruchirappalli District

It is proposed to establish an Agri Export Zone exclusively for banana in Tiruchirappalli District through private sector participation at a project cost of Rs. 10.00 crores. Necessary proposal has been sent through Government of Tamil Nadu for recommending to APEDA, Government of India for approval.

ESTABLISHMENT OF COLD CHAIN UNITS

To minimise post harvest losses of agricultural produces, especially fruits and vegetables, it is proposed to set up Cold Storage Units through public-private participation utilising of RIDF funds.

FOOD PROCESSING INDUSTRIES

Food processing is gaining momentum as food processing industries ensure steady and better price to the farming community as well as availability of the commodities in processed form to the consumer all through the year. Processing of food eliminates wastage of agricultural produce to a greater extent. By cultivation of good quality processable agricultural produce the farmers stand to gain better returns. A Food Park at Nilakkottai Industrial Estate (Dindigul District) is being set up at a total cost of Rs.13.00 crores vide GO MS NO: 23, Agriculture (AM3) Department, dated: 23.01.2004. This department as a Nodal agency for Ministry of Food Processing Industries, Government of India, processes applications for obtaining grant from the Ministry of Food Processing Industries. During the financial year 2004-05, 66 proposals worth Rs. 102.77 crores to establish food processing related projects have been forwarded to the Ministry of Food Processing Industries, Government of India for grant of Rs. 14.95 Crores.

TAMIL NADU STATE AGRICULTURAL MARKETING BOARD

The important functions of the Board are as follows:

Publicity and Propaganda

Tamil Nadu State Agricultural Marketing Board carry out publicity and propaganda work on the advantages of selling agricultural produce in Regulated Markets through four centres at Chennai, Madurai, Coimbatore and Tiruchirappalli.

Training to Farmers and Staff

The Training Centre at Salem caters to the training needs of the Market Committee employees and the farmers. Three different training programmes, viz; Graders Training, Market

Committee Employees Refresher Training and Personal Contact Programme are conducted in this Training Center.

Establishing Marketing Endowment Chair at Tamil Nadu Agricultural University

Tamil Nadu State Marketing Board has created an Endowment Chair at the Centre for Agricultural and Rural Development Studies, TNAU, Coimbatore, with a deposit fund of Rs.50.00 lakhs. Various research studies and training programmes are being conducted utilizing the interest that accrues from the fund.

Construction Works

The Engineering wing of the department executes civil works such as construction of office buildings, rural godowns, auction platform, shopping complex, agricultural input shops, payment counters, rest shed cum payment counter, water supply, toilet facilities, internal roads, godowns, compound wall, etc. in the Regulated Markets.

Market Development Fund

The Tamil Nadu State Agricultural Marketing Board derives its finance from the Market Committees. The Market Committees contribute 15% of their receipts to the Board. Fifty percent of this amount is set apart as Market Development Fund, from which expenditure towards market developmental activities including training / publicity and propaganda are met.

Establishing Domestic and Export Market Intelligence and Guidance Cell at Tamil Nadu Agricultural University

Tamil Nadu Agricultural University is in the process of establishing Agricultural Produce Domestic and Export Market Intelligence and Guidance Cell at an initial cost of Rs.20.00 lakhs from the Tamil Nadu State Agricultural Marketing Board. The cell would collect details on prices of major commodities at domestic and international markets, analyse and forecast future domestic and export prices. The cell will transmit the forecast prices in the forthcoming months to the regulated markets and farmers. The Cell by providing this information will help the farmers to plan the cropping pattern and the right time of sale of their produce. The Agricultural Production and Information centres in regulated markets will be linked to the network of above Agricultural Produce Domestic and Export Market Intelligence and Guidance Cell.

Table- 3 Market Committee Wise Regulated Markets

1. KANCHEEPURAM	2. VELLORE
1. Kancheepuram	1. Vellore
2. Tiruthani	2. Tirupathur
3. Thiruvallur	3. Arcot
4. Redhills	4. Arakonam
5. Maduranthagam	5. Vaniyampadi
6. Uthiramerur	6. Kaveripakkam
7. Ponneri	7. Gudiyatham
8. Thirukkalukundram	8. Kalavai
9. Sunguvarchatram	9. Ammoor
10. Pallipattu	10. Katpadi
11. Uthukottai	11. Ambur
12. Acharapakkam	12. Thimiri
13. Gummidipoondi	
14. Nasarethpettai	
3. THIRUVANNAMALAI	**4. CUDDALORE**
1. Tiruvannamalai	1. Virudhachalam
2. Arani	2. Cuddalore
3. Vandavasi	3. Panruti
4. Chetpet	4. Thittakudi
5. Cheyyar	5. Kattumannarkoil
6. Polur	6. Chidambaram
7. Chengam	7. Kurunchipadi
8. Pudupalayam	8. Sethiyathoppu
9. Vanapuram	9. Srimushnam
10. Vettavalam	10. Bhuvanagiri
11. Thellar	
12. Mangala Mamandoor	
13. Desur	
14. Peranamallur	
15. Dhusi	
16. Kilpennathur	

5. VILUPURAM	6. SALEM
1. Tindivanam	1. Salem
2. Tirukoilur	2. Athur
3. Ulundurpet	3. Namakkal
4. Viluppuram	4. Rasipuram
5. Chinnasalem	5. Thiruchengodu
6. Kallakkurichi	6. Sankagiri
7. Gingee	7. Konganapuram
8. Thiagadurgam	8. Kollathur
9. Sankarapuram	9. Velur
10. Tiruvennainallur	10. Mecheri
11. Manalurpet	11. Vazhapadi
12. Avalurpet	12. Thammampatti
13. Marakkanam	13. Namagiripettai
14. Vikravandi	14. Thalaivasal
15. Ananthapuram	15. Omalur
16. Valathi	16. Kadyampatti
17. Moongilthuraipattu	17. Gangavalli
18. Kilianur	18. Karumanthurai
	19. Cholakkadu

7. DHARMAPURI	8. COIMBATORE
1. Dharmapuri	1. Tiruppur
2. Krishnagiri	2. Avinashi
3. Hosur	3. Sevur
4. Kelamangalam	4. Annur
5. Palacode	5. Karamadai
6. Pennagaram	6. Coimbatore
7. Pochampalli	7. Sulur
8. Kaveripattinam	8. Palladam
9. Uthangarai	9. Udumalpet
10. Harur	10. Anaimalai
11. Pappireddipatti	11. Pollachi
12. Kolathur	12. Malayadi Palayam
13. Bargoor	13. Negamam
14. Rayakottai	14. Kinathukkadavu

15. Denkanikkottai	15. Thondamuthur
16. Papparapatti	16. Madathukkulam
	17. Pethappampatti
	18. Pongalur
9. ERODE	10. TIRUCHIRAPALLI
1. Erode	1. Jeyankondam
2. Avalpoondurai	2. Karur
3. Kodumudi	3. Ariyalur
4. Sivagiri	4. Manapparai
5. Chithode	5. Andimadam
6. Bhavani	6. Thuraiyur
7. Boothapadi	7. Perambalur
8. Anthiyur	8. Lalgudi
9. Mylampadi	9. Tiruchirapalli
10. Kavundampadi	10. Thottiyam
11. Gobi	11. Manachanallur
12. Nambiyur	12. Kulithalai
13. Vellakkoil	13. Thuvarankurichi
14. Sathiyamangalam	14. Irumputhipatti
15. Punjai Pulliyampatti	15. Chinnatharapuram
16. Thalavadi	16. Pullampadi
17. Perundurai	17. Thathaiankarpet
18. Kunnathur	18. Melanikuzhi
19. Kangeyam	19. Arumbavur
20. Vellakoil	
21. Dharapuram	
22. Moolanur	
23. Alangeyam	
24. Muthur	
25. Elumathur	
11. THANJAVUR	12. PUDUKKOTTAI
1. Athirampattinam	1. Alangudi
2. Ammapettai	2. Aranthangi
3. Budalur	3. Pudukkottai
4. Kumbakonam	4. Kandarvakkottai

5. Madukkur	5. Avudayarkoil
6. Orathanadu	6. Keeranur
7. Pattukkottai	7. Keeramangalam
8. Papanasam	8. Ponnamaravathi
9. Peravurani	9. Illuppur
10. Thanjavur	10. Karampakkudi
11. Vallam	
12. Thirupananthal	
13. Pappanadu	

13. MADURAI	14. RAMANATHAPURAM
1. Thirumangalam	1. Virudhunagar
2. Usilampatti	2. Rajapalayam
3. Melur	3. Sathur
4. Madurai	4. Arupukkottai
5. T.Kallupatti	5. Srivilliputhur
6. Vadipatti	6. Watrap
	7. Vembakkottai
	8. Sivaganga
	9. Thiruppuvanam
	10. Manamadurai
	11. Singampuneri
	12. Karaikudi
	13. Ilayankudi
	14. Devakkottai
	15. Ramanathapuram
	16. Paramakudi
	17. Kamuthi
	18. Thiruvadanai
	19. Rajasingamangalam
	20. Mudukulathur

15. TIRUNELVELI	16. KANYAKUMARI
1. Kovilpatti	1. Ethamozhi
2. Sankarankoil	2. Vadaseri
3. Thoothukudi	3. Kaliyakkavilai
4. Pudur	4. Monday Market

5. Kadambur
6. Kalugumalai
7. Thenkasi
8. Ambasamudram
9. Valliyur
10. Srivaikundam
11. Tirunelveli
12. Vilathikulam
13. Kadayanallur
14. Thisayanvilai
15. Pavurchatram
16. Thiruvenkadam
17. Ettayapuram
18. Sivagiri
19. Alangulam
20. sathankulam

5. Kulasekaram
6. Thoduvatti

17.THENI

1. Theni
2. Cumbum
3. Bodinayakkanur
4. Chinnamanur
5. Andipatti
6. Uthamapalayam
7. Periyakulam

18. DINDIGUL

1. Dindigul
2. Ottanchatram
3. Palani
4. Natham
5. Bathlagundu
6. Gopalpatti
7. Vadamadurai
8. Vedachandur

19. NAGAPATTINAM

1. Kivalur
2. Kuttalam
3. Mailaduthurai
4. Nagapattinam
5. Chembanarkoil
6. Sirkazhi
7. Vedaranayam
8. Thiruppondi

20. TI R U VAR U R

1. Valangaiman
2. Koradacheri
3. Mannarkudi
4. Poonthottam
5. Vaduvoor
6. Kudavasal
7. Thiruvarur
8. Thiruthuraipoondi

Table- 4: Districtwise Agmark Grading Laboratories In Tamilnadu

Sl. No.	Name of the District	Place of the Agmark Grading Laboratory
1	Chennai	Chennai (North)
		Chennai (South)
		Principal Lab at Office of the Commissioner of Agricultural Marketing and Agri. Business
2	Vellore	Vellore
3	Cuddalore	Panruti
4	Thanjavur	Thanjavur
5	Thiruchirappalli	Thiruchirappalli- I
		Thiruchirappalli- II
6	Karur	Karur
7	Madurai	Madurai - 1
		Madurai - 2
8	Theni	Theni
9	Dindigul	Dindigul
10	Virudhunagar	Virudhunagar
11	Tirunelveli	Tirunelveli
		Thenkasi
12	Thoothukudi	Thoothukudi
13	Kanyakumari	Nagarkoil
		Marthandam
14	Salem	Salem
15	Dharmapuri	Dharmapuri
16	Coimbatore	Coimbatore
		Tiruppur
		Palladam
17	Erode	Perundurai
		Erode - I
		Erode - II
		Kangeyam - I
		Kangeyam - II
		Vellakovil
		Chithode

STATE ASSISTED AGRICULTURAL ORGANISATIONS

Tamil Nadu Horticultural Producers Cooperative Enterprises Limited (TANHOPE)

Tamil Nadu Horticultural Producers Co-operative Enterprises Limited (TANHOPE) was registered in the year 1995 as a primary Co-operative society under the Tamil Nadu Co-operative Societies Act 1983 and started functioning from 18.1.1995 onwards.

The main objectives of this institution are to encourage and promote horticultural activities with good market linkages in the State Tamil Nadu. The area of operation of the institution is the entire State of Tamil Nadu. The Director of Horticulture and Plantation Crops is its Special Officer and Joint Director of Horticulture is the Managing Director.

As on date, this institution has enrolled 3888 grower members from various districts with a paid up share capital of Rs.3.88 lakhs.

The annual business turn over of this institution of the past three years are as follows.

- 2001-2002 - Rs.38.97 lakhs
- 2002-2003 - Rs.79.48 lakhs (inclusive of 50.28 lakhs Cashew grafts supplied to Horticulture Department)
- 2003-2004 - Rs.31.06 lakhs
- 2004-2005 - Rs.22.78 Lakhs (Upto February 2005)

It is proposed to facilitate marketing of produces of the Self Help Groups since cultivation of horticulture crops is presently promoted as a group based activity in selected districts.

During 2005-2006, TANHOPE will focus in strengthening the group based horticulture activities with special emphasis on value-added products, post-harvest management activities and marketing of produces.

12

Researching the Drylands

Â'Hand over the Ministry of Agriculture to a politician from the arid Deccan plateau; someone who has known hunger, distress and seasonal migration. And make a political commitment to revamp agricultural science and markets in these regions.Â' This was the reply from an eminent scientist, in 1991, when Prime Minister Narasimha Rao asked him what he could do Â'to transform dryland agriculture and bring prosperity to poor villages spread over ten states.Â' In the words of this eminent scientist, the meeting ended on a sad and helpless note. Â'Our agricultural research system[1] has become too rigid and set in its ways. Even if markets change, we in the sciences may not change. We are blinded by our green revolution euphoria. We can no longer see reality.Â' Little has changed since then. Highlighting the bleakness is the ICARÂ's Vision 2020 Â– the absence of the word Â'droughtÂ' and an exclusively crop-based strategy to address rainfed agriculture![2]

This essay explores the inability of agricultural S&T to understand and respond to the demands of dryland agriculture. Though there are several problems that are generic to agricultural science and technology, some of these are specific obstacles to the generation and utilization of relevant knowledge for the resource-constrained agriculture of the drylands. If scientific research is to make a significant contribution to dryland agriculture, there must be (*i*) a broader and more inclusive definition of knowledge and issues relevant to livelihoods in the drylands, and (*ii*) a new paradigm of institutional learning and change, where science and technology will become part of a wider coalition of actors, inputs and processes. To ensure sustainable rural livelihoods, the public sector, in partnership with other actors where feasible, has to launch a new knowledge initiative for the drylands along with other investments in infrastructure, natural resource management and human resources development. This is as much a political challenge as a technological one.

Historically, formal scientific research for dryland agriculture was organised to serve the colonial interest in dryland crops of commercial value Â– cotton, indigo, sugarcane, wheat, lac, and so on. Whether it was experimental research in the country or transfer of technology (especially varieties and new implements) from the West to India, these efforts were conducted by individual researchers or in experimental farms funded by the government or by Commodity Boards using a specific fund (based on a cess levied) for research. Some of the early successes in rainfed rice, sugarcane and cotton came from these experimental farms. All these different research organizations under the provincial governments, commodity boards, quasi-public/private farms, etc. were taken over and consolidated within the ICAR in 1965.

Contrary to popular perception, dryland agriculture in India even today is marked by a variety of actors/organizations and processes. In the public sector, the ICAR institutes and the State Agricultural Universities conduct research for the drylands. The ICAR has specific research organizations Â– the Central Arid Zone Research Institute (CAZRI), the Central Research Institute for Dryland Agriculture (CRIDA), and parts of the Indian Grasslands and Fodder Research Institute (IGFRI), Central Soil Salinity Research Institute (CSSRI) mandated to conduct research on specific crops and problems in the drylands. The International Crops Research Institute for the Semi Arid Tropics (ICRISAT) is located in India, and several donor-funded programmes focus on research for dryland agriculture (such as the Aga Khan Rural Support Programme in Gujarat, or DFIDÂ's programmes in the western and eastern regions).

Private and cooperative sectors are also active in R&D in the dryland farming tracts, especially in sectors such as private plant breeding and seed production. Over 40 private firms breed and sell certified seeds of dryland crops (especially cotton, pulses, oilseeds, maize and sorghum). Some of the cooperatives have been very successful in dryland production systems like milk (Amul and other state dairy cooperatives), sugarcane, grapes, and other fruits. The cooperatives, more adept at collecting and adapting technologies and information for their own farming community members (than in conducting research/ technical experiments), have proven how different ways of communication and delivering information can enable better technology adoption.

Farmer-to-farmer extension is a major source of technological information as well as new technologies (seeds/ crop varieties, pesticides, crop protection practices, irrigation practices etc.) in the drylands. Traditional wisdom and local cultural knowledge seems to be critical for survival in the drylands.

Despite the involvement of several national and international players and organisations, the overall performance of formal R&D for dryland agriculture has been far from satisfactory. R&D investment has been low, and the use of that investment (as we shall discuss here) has been highly ineffective for dryland agriculture. Measures such as research intensity (R&D investment in rupees per unit of commodity/crop produced) are misleading. They show that dryland crops (pulses and oilseeds, and some coarse cereals) fare better than irrigated crops (especially wheat, rice, and sugarcane) (Kerr, 1996). But this is clearly because of the massive production and productivity differences between dryland crops and irrigated crops. Nor do these measures give us an adequate picture of the research effort that needs to go into dryland agriculture. Moreover, measuring research intensity adds to the myth that increased investment within existing research organizations and paradigms is in itself a solution.

The arid and semi-arid tracts in India contribute over 45% of the national agricultural production. Rainfed agriculture in 100 million hectares of arable land is the manifest opposite of the prosperous irrigated cereal-based green revolution tracts.

It is accepted that the green revolution has bypassed millions living in the drylands, surviving on one rainfed crop a year Â-- which is bound to fail or is not taken at all if the monsoon is poor, erratic, delayed or absent (ICAR, 1998).

The ICARÂ's classification of the country into 20 agro-climatic zones includes rainfall, soil type and length of growing period as the defining features of each zone. Using rainfall and access to irrigation as defining features of dryland agriculture, all agro-climatic zones with the exception of five zones, are either completely or partially dryland agriculture zones. The need for a different typology to develop dryland/rainfed agriculture using S&T inputs becomes obvious considering the range of complexities associated with dryland agriculture that extend beyond agro-climatic conditions and include socio-economic realities of these tracts.

Dryland agriculture is marked by risk and coping strategies to avoid risk. These strategies include diverse crops and cropping systems, dependence on livestock and other non-farm rural income, limited cash crop production (where assured markets exist), common property based livelihoods or access to livelihoods, low input use and technology adoption. Household income in a typical arid/semi-arid village is subject to a high degree of seasonal variation. Seasonal out-migration, lasting four to six months every year, is routine. These villages are also marked by relatively high proportion of scheduled caste and scheduled tribe households (traditionally disadvantaged to access modern input intensive agricultural technology), low or negligible rural infrastructure investments and services, high incidence of morbidity and malnutrition, illiteracy and limited access to education, all revealing multiple dimensions of poverty. In terms of crop production the rainfed cropping systems also reveal very high inter-farm variation within the same village.[3]

Thus, dryland agriculture in India is marked by several constraints (technological and non-technological) and for a full understanding of these, one needs to go beyond the narrow technical criteria usually adopted to define drylands.

IndiaÂ's experience with generating technological solutions for technological problems reinforces the definition of dryland agriculture in narrow technical terms. The green revolution is premised on the definition of hunger and malnutrition as food production problems. Thus the national goal of food self-sufficiency is translated into an applied research strategy to produce technological solutions. Risk proofing through assured irrigation, chemicals and fertilizers, and high yielding varieties was the hallmark of the green revolution technologies. Applied research within the key disciplines now focuses almost exclusively on technology generation to assure this green revolution paradigm Â– albeit within the narrow range of controlled conditions.

Agricultural research addresses the problems of dryland agriculture assuming that the same norms of irrigated agriculture can be applied here. Attempts to understand dryland agriculture and solutions offered are always technology centred Â– crop-based (varieties), input-based and irrigation based (water use efficiency, water saving technologies, etc.).

Several scholars lament the inadequate attention paid within the crop sciences to the rainfed crops Â– especially the more stress tolerant and nutritious coarse cereals and minor millets/pulses/oilseeds. Even in cases where such attention to coarse cereals is given, varieties are screened in multi-location trials and assessed using average yield/productivity criteria from test plots across the country. This means that varieties released are often not ones that survive the extreme stress levels of the drylands. It also means that the local adaptation levels of the varieties are not a concern when assessing and releasing crop varieties Â– though inter-farm variability in yields even within one village is a major issue in the drylands.

Farmers in the dry tracts look for higher yields, but as one among many other traits like grain to fodder ratio, crop duration, seed quality, drought tolerance and pest resistance. The rainfed rural economy has adapted to the inherent instability in crop yields by cultivating diverse crops, engaging in several livestock based and other non-farm options. Often fodder yield is considered more crucial than grain yield Â– especially in the dry villages subject to severe seasonal stress.[4] Plant breeding as a discipline has limited capacity to understand the varietal selection and livelihood options in the drylands. Even soil science research is assessed using the criterion of yield enhancement; the knowledge inputs and conservation/resource management services from soil science notwithstanding. Given the lack of assured irrigation, fertilizer use and other chemical/pesticide application is limited in the drylands. Besides lack of information about use and availability, there is also a question of access to these inputs (including access to credit and repayment capacity estimated against incremental income even if the technology is for medium or long term natural resource management). Agricultural science knows precious little about the choices that farmers make for land management in the drylands Â– even share cropping (in villages where entire families migrate for over four-six months every year) or crop-livestock relationships are analysed using the same technical and economic criteria that are used in irrigated areas.

Agricultural science has limited capacity to conduct experiments along the scales that determine input use and resource management in a dryland farm. Watershed development activities have shown that households in a village can be made aware of the layout of a watershed and availability of groundwater and appropriateness/impacts of input use at different points in the watershed. But the soil sciences lack experimental designs (and partnerships with local stakeholders) to integrate technical information to assess soil quality (capacity for nitrification, porosity, and other structural features), along the watershed contours.

As a rule, agricultural research is capable only of plot cultures or plot level on-station/on-farm experiments (even for ritualistic fertilizer trials).[5] Even in dryland R&D organizations with some on-farm research in addition to the on-station research (CRIDA for instance), the emphasis is on taking on-station research design or results to the farm. There is no attempt to learn with the farmer or to understand local topographic or edaphic factors. The lack of communication and learning across disciplines within the agricultural sciences is an affliction in dryland agricultural R&D too. Communication among sub-disciplines of the same discipline (say the sub-disciplines of soil science) is minimal or non-existent. Dryland agriculture is defined by lack of assured irrigation; yet scientific research for the drylands is governed by the norms of irrigated agriculture. R&D assumes that enhancing irrigation in the drylands will solve all the problems. The availability of water and the diversity of water and land use patterns (be it the absence of irrigation in the arid Rajasthan tracts that produce high grade *mehndi* for export, the tank irrigated rice production systems in South India, or the fast-depleting-groundwater-irrigated groundnut and sugarcane systems of Gujarat/Maharashtra/western Madhya Pradesh) ought to have conveyed the message to agricultural science that irrigation and green revolution type production enhancement technologies are not the answer to dryland agriculture. Even if all the irrigation potential is used, 40% of IndiaÂ's food requirements will still need to come from the dryland tracts.

The recent spate of farmerÂ's suicides in these areas proves how the green revolution package dependent on high levels of water use, purchased inputs, and reliable markets does not even comprehend the limited resource availability and high inter and intra-seasonal variability that characterize dryland agriculture. Sadly, rural development schemes and private investments that encourage the exploitation of groundwater for irrigation of water intensive cash crops in the drylands, rely on Â'technology packagesÂ' specially developed by agricultural research organizations, and fed into these schemes by subject matter specialists.

Water management, especially integrated water resources management is the answer for the dryland tracts. Successful cases of NGO or other civil society led watershed management programmes reveal how farmers and other water users in a watershed can agree about and effectively manage their water resources and crop/livestock production systems through appropriate processes, investment strategies, benefit sharing mechanisms, conservation technologies, production technologies, etc. The distinguishing feature here is the absence of a single point Â'technologicalÂ' solution as found in irrigated crop production. Here technology, be it irrigation development or plant protection methods, becomes part of a more comprehensive package of processes and inputs.

The agricultural sciences in India are so organized as to insulate its disciplines from learning lessons from the field in collaboration with (*a*) other disciplines Â–social sciences, natural resources research, animal husbandry, among others, or (*b*) other non-research partners Â– civil society organizations,line departments, private firms, financial organizations and farmers organizations. Opportunities for interdisciplinary research and working in collaboration with other partners in the drylands declined rapidly following the consolidation and centralization of public sector agricultural research Â– especially given the norms for evaluation of research based on peer reviewed research publications.

Dryland agricultural research is not accountable to a local constituency Â– be it research in the public sector or private sector. Only the cooperatives seem to have made a mark in agricultural knowledge systems in the drylands. Will the future of dryland agricultural research bring new organizational formats and new partners? The emerging message is one of inadequacy of the existing paradigm of R&D to address the knowledge and development needs of the drylands.

The approach of delivering technological solutions to development problems that are chiselled down into narrow technical problems (using convenient criteria) is universal. In the case of dryland agriculture, even the scientific solution itself seems to be limited by the narrowly defined Â'technicalÂ' criteria (of rainfall, soil type or growing period) leaving out any reference to other ecological (status of water table/land degradation Â– erosion, salinity) or socio-economic criteria. Conventional neo-classical agricultural economics and extension science in the research organizations is structured in such a way as to impede the entry of new paradigms that bring better understanding of social contexts and have sustained the conventional linear R&D models and technology generation functions in R&D organizations. At this current juncture, dryland agriculture demands that the social sciences lead agricultural R&D by helping the technical sciences understand rural realities in these tracts. There are several examples of agricultural and rural innovation (defined as the socially embedded utilization of technologies and other knowledge) that offer important lessons for the agricultural sciences in the drylands. Agricultural research should:

(a) Shift mandates from research to innovation: Given the complexity of production conditions in dryland agriculture, the emphasis has to increasingly shift towards utilization of knowledge in economically and socially productive ways. To ensure successful innovation, the researchers and research organizations should be encouraged to find appropriate coalitions to work in, with local accountabilities and mandates designed to serve local contexts. For them research output or even the choice of research problem should become one among several steps in a flexible innovation programme, to achieve the desired development outcome.

(b) Improve its capacity to learn from local contexts: The shift of emphasis towards innovation presupposes a change in the attitude of researchers and research organisations towards local knowledge. They should have the humility to learn from local contexts before attempting to intervene in them.

Agricultural science needs to understand that agriculture is one among several sources of livelihoods for rural households in the drylands. Unless these other livelihoods options, their relationship with agriculture, local contexts and systems are understood, and a diverse range of stakeholders involved in innovation, technologies and institutional arrangements necessary for development cannot be identified and enabled. The attitude of learning from local contexts also should make the researchers mindful of the multiple impacts of their actions on the local environment.

(c) Build new relationships and collaborations with other actors: All the cases of successful innovation and development evident in the drylands (cheap drip irrigation systems, integrated pest management, watershed development, etc.) reveal several actors working synergistically as an interested and committed coalition, improvising and nuancing their interventions with a deep respect for local knowledge Â– whether it is from input dealers, industry, extension staff/ departments, or illiterate risk averse farmers.

The foregoing discussion clearly shows that there is an urgent need to change the rules/norms governing the generation and application of knowledge in the drylands. The paradigm of linear knowledge flows and technological determinism has to be changed and the new norms/institutions for a learning community need to be laid out. More funds or new organizations can help dryland agriculture only if accompanied by such a change in approach and in the rules of the game.

The answer presently favoured is to privatize everything Â– presumably because reforming failed public sector is considered a hopeless task. There are powerful forces at work, which attempt to transfer research and extension from public to private/voluntary agencies. While accepting that increasing collaboration among public, private and civil society actors is important, the key role of the public sector in research and extension needs to be re-emphasised. In fact, the public sector has to play several roles besides generating technology. It has to play a dynamic monitoring and planning role to get these collaborations to work in a resource poor dry-land context. It also has to ensure that knowledge is used as part of a wider package of inputs and processes for sustainable development in drylands, with a great deal of flexibility in ways of working and choice of partners in different contexts.

RAINFED AREAS

Rainfed regions are those where crop production is exclusively dependent upon rainfall. In India rainfed regions cover 177 districts and exist in all agro-climatic zones. However, they are mostly concentrated in arid and semi-arid areas. Most of these districts are country's poorest. Rainfed regions account for 68 per cent of the total net sown area in the country, according to the Union Ministry of Agriculture.

An assessment on agriculture done by S M Jharwal, principal advisor to the government of India, shows that out of a net sown area of 141 million hectares (ha), 86 million ha is rainfed. Of the 190 million ha cropped area, the ultimate irrigation potential is assessed at 140 million ha. This potential takes into account major, medium, and minor irrigation systems, using surface as well as ground water sources.

Already, the irrigation potential of about 103 million ha has been created, but the actual gross irrigated area is only 77 million ha. Based on these assessments, it emerges that currently 113 million ha area is rainfed, and even if we create a full irrigation potential of 140 million ha, based on the current utilisation (75 per cent), about 85 million ha would remain rainfed.

However, rainfed areas change depending on rainfall and water availability in reservoirs. During 1993- 94 and 2003–2004, rainfed areas declined from 90.88 million ha to 85.78 million ha.

State-wise assessment shows that 13 states account for about 92 per cent of the total rainfed area. These include the main states of Maharashtra (14.49 million ha), Madhya Pradesh (9.31 million ha), Rajasthan (12.15 million ha), Karnataka (7.46 million ha), Uttar Pradesh (4.42 million ha), Andhra Pradesh (6.48 million ha), Gujarat (6.58 million ha) and West Bengal (2.54 million ha).

Fragile, Degraded

Ecologically, rainfed areas are the most fragile even though they sustain substantial populations. One-third of the dryland areas are highly degraded, which cannot be put under cultivation. They receive rainfall less than 500 mm or more than 1,500 mm and suffer from serious water management problems either way.

Some of these areas have black cotton soil, which has its own problems. Around 40 per cent of the estimated 100 million ha of dryland areas have shallow depth and have been affected by massive soil erosion limiting their production potential. Rainfed regions are highly drought-prone. On an average, the rainfed regions of India suffer from droughts once in every three years. Western Rajasthan, eastern Rajasthan, Gujarat, western Uttar Pradesh, Tamil Nadu, Kashmir and Andhra Pradesh are most vulnerable to droughts.

Often the drought persists for three to six years at different intensities causing multiplier negative effects on availability of water for drinking to humans and livestock, and on crop and fodder production.

80 per cent Rainfed?

Going by the literary meaning of 'rainfed', (see box: It's not irrigated agriculture), many more areas in India can be termed rainfed. Of the 54.56 million ha net irrigated area, 69.7 per cent is dependent on tanks, wells, and traditional water bodies. Barring the wells under command area, the rest, along with other sources, are also dependent on rain. This means that around 80 per cent the areas are rainfed.

According to the Central Ground Water Board, in 100 districts of India excessive ground water use has led to major economic and sustainability problems. These districts are termed 'critical' as far as their ground water scenario is concerned.

"These 100 districts account for over 60 per cent of India's 'critical' and over exploited blocks. Eighty-five of these districts are in the rainfed regions. These areas happen to have the highest concentration of dug-wells in the country. Here falling water tables have the most disastrous impact on drying wells, forcing farmers to revert to rainfed farming," says former planning minister Y C Alagh.

In fact, the working group on rainfed farming constituted by the Planning Commission to prepare the 11th Five-Year Plan said that the areas where ground water is stressed, should also be treated as rainfed areas and be covered under related programmes. This is because ground water is recharged by surface rainwater.

India's irrigation strategy has mostly focused on surface water use even though groundwater use has increased drastically, particularly in rainfed areas. The expenditure on surface irrigation during the 10th Five-Year Plan was

Rs 95,734.4 crore, compared to Rs 441.8 crore in the first plan. During the first four five year plans, India created huge projects to harness surface water. Irrigation picked up. However, there was utilisation of available land and water. During the 5th Five-Year Plan (1974-78), the Command Area Development Programme was established to bridge this gap.

Immediately after this, new projects were again opened up for the 6th Five-Year Plan (1980-85). At the end of the 7th Five-Year Plan, there were 182 major and 312 medium projects requiring an estimated Rs 39,044 crore for completion.

The accelerated Irrigation Benefit Programme was launched in 1996-97 for this purpose. This created irrigation for 93.5 million ha land by the end of the 9th Plan.

To illustrate, let's take the example of 99 drought prone districts in 14 states. These districts have 42 per cent of the country's cultivable lands and the government has 148 major and 195 medium projects. Most of them have not been completed.

So, the irrigated lands in these districts are what they used to be in the 1960s. In the 11th Five-Year Plan, the government proposes to bring 16 million ha under the irrigation potential with a target of bringing at least 64 million ha of new area under irrigation. Meanwhile, some 72,000 tanks and storages for rainwater are not being used.

What do Rainfed Areas Mean for India's Agricultural Sector?

Rainfed agriculture plays an important role in India's economy. Rainfed crops account for 48 per cent of the total area under food crops and 68 per cent of the area under non-food crops in the country. Nearly 50 per cent of the total rural workforce and 60 per cent of the livestock in the country are concentrated in the dry districts. "Rainfed areas are a paradox. On the one hand they remain where they were... but they also sustain a vast population in the most difficult situations," says S P Mittal, former Principal Scientist at the Central Soil and Water Conservation Research and Training Institute.

Further, crop-wise analysis shows that major coarse cereals are grown in rainfed areas. Coarse cereals are still the main source of food for India's poor. For instance, 92 per cent, 94 per cent and 80 per cent of the total area under Jowar, Bajra, and Maize respectively is rainfed.

Similarly, 86 per cent of the area under pulses is rainfed. Eighty three per cent groundnut and 99 per cent soybean are grown under rainfed conditions. About 73 per cent area under cotton is rainfed.

Though rainfed areas contribute in a major way to India's agriculture, the difference between the output of rainfed and irrigated areas is remarkable. This is cited as a major reason for increasing regional disparity in India.

In fact, states with large rainfed areas have reported less agricultural growth in the last decade (see: Growth rate in irrigated regions). On May 29, 2007, during the National Development Council meeting to finalise the 11th Five-Year Plan, Prime Minister Manmohan Singh clearly stated this difference as a reason for worry as well as the target for the next Green Revolution he has been championing. "There is a technology fatigue in Indian agriculture," Singh said, pointing to the Green Revolution method of increasing productivity.

A redeeming feature of the semi-arid tropic regions is that there is a wide variation in crop yields within the region. For example, the average yield of sorghum varied from 370 kg/ha in farmers' fields to 1,460 kg/ha in demonstration fields and 1,060 kg/ha with assured input supply.

There is still considerable scope for increasing productivity in fields because the yields actually being achieved for many crops in many states are much below their realisable potential (even of existing varieties if best cultivation practices are used).

In wheat, the unutilised potential is only 6 per cent in Punjab, but rises to 84 per cent in Madhya Pradesh. In Maize, the gap is only 7 per cent in Gujarat, but 300 per cent in Assam. In rice, the potential yield increase is over 100 per cent in Assam, Bihar, Chhattisgarh and Uttar Pradesh.

This is reflected in the fact that the yield of new varieties released in recent years has plateaued. Past strategies, which ushered in the Green Revolution in the 1970s and 1980s, are not working anymore. Thus, there is a need to identify and focus on new strategies to achieve these goals.

In areas that witnessed the Green Revolution, the productivity levels are high. However, over the past decade yields have been stagnating and in some cases even declining.

Past resources of growth productivity like expansion in irrigation, and increased use of fertilisers and chemicals for pest control are no longer relevant. Policy regimes, which helped achieve increased productivity, are now not only irrelevant but are also contributing negatively to resource quality.

WHY SHOULD WE FOCUS ON RAINFED AREAS?

India's Key to Food Security

As opportunities for further agricultural growth in irrigated regions get exhausted, food security and productivity growth in agriculture in India in the coming years will increasingly depend on improved utilisation of natural, human, financial and material resources and productivity growth in rainfed regions.

According to the national agriculture policy, India must achieve a growth rate of 3- 4 per cent per annum in the sector. The Union ministry of agriculture estimates that the total food grain production during 2006-07 was 209.2 million tonnes.

For maintaining food security even at the current nutritional levels, about 100 million tonnes of food grains need to be produced additionally by 2020. The total cropped area in India has remained static at around 140 million ha since 1970s. So these areas do not serve as a source of increased output. Thus, the increased yield must come from areas with the least irrigation potential. Even then, there is likely to be a shortfall of 38 million tonnes of food grain in 2020.

Further, the total contribution of irrigated agriculture to food grain production from both area expansion and yield improvement put together is likely be around 64 million tonnes in 2020 — leaving a shortfall of 38 million tonnes. (see table: Projected demand and supply of food grain in India 2020 (million tonnes)

Projected demand and supply of food grain in India in 2020 (in million tonnes)

Projected food demand in 2020	307
Average food production in triennium ending 2002	205
Gap to be met	102
Maximum possible contribution of irrigated agriculture of which from	38
• Irrigated area expansion	
• Increase in productivity of irrigated agriculture	26
Minimum balance required from rainfed agriculture	38
Share of rainfed agriculture	37 per cent

Which implies that even in the best possible scenario of irrigation development, about 40 per cent of the additional supply of food grain needed to match future rise in demand will have to come from the unirrigated segment of Indian agriculture, most of which is located in dryland areas. And this demands that productivity of drylands be raised through intensive watershed work.

According to research done by the National Bank for Agriculture and Rural Development (NABARD) on the rice economy of India, rainfed areas, particularly of eastern India, will be crucial in ensuring rice production that contributes 60 per cent of food grain production in the country.

Rice production registered a four-fold increase during 1950-90, but started declining since 1990. The decline was not only in the Green Revolution states like Punjab and Haryana, but also in rainfed areas like Andhra Pradesh and Kerala.

The NABARD research shows that increase in rice production has been skewed. "The gain due to modern rice technology has been discriminatory against the resource poor areas, which are also dominated by small and marginal farmers (read rainfed areas)," says the report, Dynamics of rice economy in India: Emerging science and policy options.

As many Green Revolution states like Punjab are resorting to crop diversification, India has to look for areas where it can compensate the decline in rice production. It is estimated that by 2010, we have to compensate around 5.5 million tonnes of rice in the process of diversification.

"Apparently in order to compensate the loss of total production in such an event, the onus automatically falls on the lesser performing rainfed areas. It would require at least three times additional area at the given level of productivity of rice or to increase the productivity itself by two to three times," says the report.

And this effort has to be made by the eastern states as these have hardly 36 per cent of the area under irrigation.

"Thus, it is the rainfed parts of Indian agriculture that have been the weakest, they are also the ones that contain the greatest unutilised potential for growth, and need to be developed if food security demands of 2020 are to be met," says S Parthasarthy, who headed the technical committee on watershed programmes in India.

CHANGE IN CROPPING PATTERN

There has been an enormous shift in crops and cropping patterns in rainfed areas. Commercial crops like sunflower, soybean, and groundnuts have replaced the rainfed staple coarse cereals. More recently cotton is replacing sorghum in its potential areas.

Mixed cropping, which was universally practiced in rainfed areas, is now limited to hot arid and humid regions besides tribal regions. With the popularisation of bore wells in rainfed areas, rice and horticultural crops like fruits and vegetable have come up. Thus, ecological access to food has become acute.

In well-irrigated rainfed ecosystems, food habits have changed enormously — rice has replaced coarse cereals. In fact, in the process the marketability of coarse cereals has become a problem. Even though the minimum support price (MSP) exists for these crops, there is no attempt to procure them in many areas.

"MSP is hardly implemented in a few states. For coarse cereals it almost doesn't exist. This makes the market, and incentive for rainfed crops unsuitable," says Ramesh Chand of the National Centre for Agricultural Economics and Policy Research.

That rainfed farmers are abandoning their most suitable crops is evident from the fact that the output of crops grown and eaten by the poorest of the poor (coarse grain, pulses and oilseeds) and grown largely in drylands, actually declined from 1995-2005 and the rate of growth of their yields decelerated considerably.

According to the working group on rainfed areas for the planning of the 11th Five-Year Plan, production of coarse grains did not show any significant trend from 1980-2003 in the country as well as in rainfed regions.

The moderate growth in productivity levels couldn't neutralise the delivering trend in production caused by the shrinking area. In case of millets, the productivity enhancement for rainfed areas was 0.29 per cent, whereas the area in rainfed region reduced by 2.17 per cent, resulting in an overall decline in production from rainfed regions by about 2 per cent.

In the use of oilseeds, though there is a decline (0.56 per cent) in production at the national level, productivity enhancement (0.75 per cent) and increase in oilseed area (3.10 per cent) under rainfed conditions resulted in increased production (3.87 per cent) from rainfed areas.

A comparison of rate of growth of production from 1980-90 and 1990-2003 indicates lower growth rates during 1990-2003 in case of coarse grains (1.91 per cent and 1.47 per cent) and millets (3.31 per cent and 0.29 per cent) and marginally higher growth rates for pulses (0.32 per cent and 0.39 per cent) and oilseeds (0.70 per cent and 0.75 per cent).

During 1980-2003, the area sown to pulses declined marginally in the country though it increased at an annual rate of 2.7 per cent in rainfed regions. However, the growth in productivity was low at about 1.2 per cent in all India and rainfed regions. Compared to the 1980s, growth in area and production slowed down considerably in rainfed regions after 1990. The growth in productivity was less than 1 per cent during this period.

The production of oilseeds increased in rainfed regions at a much faster rate compared to the all India growth during 1980-2003. The growth was however more rapid during the 1980s largely because of the introduction of technology mission on oilseeds during this period. As a result, oilseeds gained area at a rate of 8 per cent. However, this was not accompanied by yield gains. After 1990, area expansion stopped. In fact there was a small loss of area to other crops in rainfed regions.

The growth rate in productivity in rainfed regions never exceeded 1 per cent. It is now admitted that closing the oilseed technology mission in 1995 hit rainfed areas hard, as this constituted close to 60 per cent of outputs from rainfed areas.

SUITABLE APPROACHES TO RAINFED FARMING

Anantapur district in Andhra Pradesh is an example of how rainfed farmers are being pushed to greater misery due to government policies. The district has Andhra Pradesh's largest chunk of dryland farms as well as the largest areas under groundnut cultivation.

The district first took up extensive groundnut cultivation protected by the oilseed technology mission and a protected market. In the 1980s the state government gave people rice at Rs 2 a kg, thus forcing them to consume rice and abandon their subsistence mixed cropping.

Then in 1990 the oil market was deregulated and palm oil was allowed to be imported. Meanwhile, the state government had also discontinued its populist rice scheme due to the mounting burden on the state exchequer. And all of a sudden, the farmers were left high and dry.

Moreover, groundnut cultivation was becoming expensive due to inputs in the form of pesticides and fertilisers. Gradually, the yields began to drop: yield from 1 ha came down from 2,000 kg in the mid 1980s to 200 kg in 2002. Earnings from groundnut also crashed to Rs 300-Rs 500 in 1997-98, one-tenth the price just a few years ago.

There was reduction in the area under cotton during the 1980s across the country and in rainfed regions. The productivity growth was much higher (4.04 per cent) in the country compared to rainfed regions (1.94 per cent).

After the 1990s, production in rainfed regions and in the country as a whole showed a declining trend inspite of expansion of area under cotton. Take the case of Bt cotton. It might be good in certain situations — high rainfall and irrigated areas, and with farmers with high investments and risk bearing capabilities. But with universalisation in the recommendations and with greater expectations, even poorly endowed areas were put under Bt cotton.

Normally, poorly endowed areas get low rainfall and are owned by the poorer sections of society. Even in better rainfall areas the poor own lands where the uplands are light and shallow. The critical phonological stages for cotton include square formation, flower initiation and boll formation when moisture availability is necessary for obtaining normal yields. Since the poor soils/rainfall areas are more affected with moisture stress, Bt cotton should not have been chosen.

Similarly, with the traditional rice system being phased out of the landscape and being replaced with the hybrid variety, rainfed areas registered very low productivity. About 36 per cent of the districts covering 44 per cent of the total rice area in the country achieved a productivity level of more than 2 tonnes/ha.

These districts contributed the lion's share of 63 per cent of production. On the contrary, the remaining 64 districts covering 57 per cent of the total rice area had to contend with dispassionately low productivity of less than 1 tonne/ha. Most of these districts are rainfed. Rice contributes close to 60 per cent of an average person's calorie intake. This means in rainfed areas low productivity and production also affect the overall health of the people.

Studies show that the maximum increase in rice production and productivity has occurred due to three factors: the high yield variety, fertilisers and irrigation facility. Thus rapid adoption of modern variety took place in irrigated areas, which suited the production conditions using high yield varieties of rice. Thus, rainfed areas, by default non-irrigated, lagged behind. This created a crisis as farmers used the same varieties without irrigation and therefore faced uncertainty in crop outputs. An interesting picture emerged from the analysis: the proportion of area under modern varieties was found to be larger than that of area irrigated. This implied that modern varieties spread to non-targeted rainfed ecosystems also, as farmers were willing to encash the opportunities for increasing rice production through modern varieties without adequate irrigation facilities. Thus despite irrigation, production has been modest.

The low productivity in rainfed areas and ecological diversity created enormous demand for targeted technologies, particularly for ecosystem-based rice research. But the existing technology is far from adequate in rainfed areas.

More than 55 per cent of the total number of modern varieties are targeted for irrigated areas, while the same is 27 per cent in rainfed lowlands and 19 percent for rainfed uplands. This means that there are 25 modern varieties per million hectares of irrigated rice areas, only 10 in rainfed lowland and deep-water ecosystem, and 13 in rainfed upland rice ecosystem.

Due to scarcity of ecosystem specific improved varieties, the rate of adoption of modern varieties has been moderately low, and thereby the low productivity.

Let us consider maize. The area under maize is growing enormously. In fact in the recent past, it is the only cereal that has shown a growth rate of more than 2 per cent as compared to other cereals which are about 1 per cent or less.

So even very poorly endowed and low rainfall areas are put under maize, for example south Telengana. Such areas are not suitable for the maize crop. They are better suited for cereals like local varieties of sorghum.

Naturally, the maize grown in such areas could be subject to more moisture stress, leading to lower yields and poor quality grain, and unable to bear drought conditions.

"One of the serious negative effects of mono-cultural cropping pattern promoted by the Green Revolution technology has been the loss of soil fertility and productivity, undermining long-term sustainability of yield growth," says B Venkateswarlu, head, Division of Crop Sciences, Central Research Institute for Dryland Agriculture, a Hyderabad-based institute under Indian Council of Agriculture Research (ICAR).

Agricultural scientists point out that there is a need to address the declining productivity of land in rainfed areas. It is estimated that in the last 15 years, there has been a deceleration in the growth rate in all sectors such as crops, livestock, fish etc. This is largely due to continuous loss in soil productivity.

Another worrying factor is the loss in soil organic matter (SOM), which leads to deteriorating physical and biological conditions. Estimates say that annually crops take 28 metric tonnes of nutrients away, and only 18 metric tonnes of chemical nutrients are added in the country. Out of the chemical nutrients around 10 metric tonnes are used in the production systems, the rest being lost into the atmosphere and soil, thus polluting the biosphere.

"This is happening because of the government's subsidy for fertilisers," says J Venkateswarlu, former director of Jodpur-based Central Arid Zone Research Institute (CAZRI), under the ministry of agriculture.

Evidently, the rest of 18 metric tonnes of nutrients are derived from natural resources. Thus there is a great depletion of SOM, which is close to 0.2 per cent to 0.4 per cent in different ecosystems. So there is an urgent need to encourage use of organic nutrients and enhance SOM. "Soil is a living system and it has to be treated differently. Due to low soil organic matter, the water absorbing capacities of soil have gone down," says Venkateswarlu.

Where are the 'Treated' Areas?

What happened to the extensive rainfed areas treatment plan? By the end of the 10th plan, nearly 51 million ha were supposed to have been "treated" under watersheds jointly by the ministries of rural development, agriculture and environment and forests, and by the Planning Commission for nearly Rs 20,000 crore. The 11th plan is hoping to treat additional 38 million ha, costing about Rs 28,000 crore. Although positive outcomes of watershed-based developments are reported from several locations, little is known of the impact of the past "treatments" on national production, productivity, farmers' income, and equity indicators.

Cases of reversal to the "untreated" original situation are commonly encountered. Obviously, the "business as usual" will not do. The one-time project-based treatment approach under watersheds is unsustainable. "The watershed plus approach has not been fully implemented. The programmes/subsidies supporting agriculture are not differentiated to suit the requirements of rainfed farming. Support in terms of fertiliser subsidies, price support and procurement does not reach in the same way for rainfed farmers, resulting in skewed national investments across irrigated and rainfed areas, says the Parthasarathy Committee report. The historical neglect and absence of appropriate support systems have created a high degree of indifference on part of farmers towards rainfed agriculture. Private investment and care for natural resources have suffered. Natural resources degradation cannot be arrested if farmers are apathetic towards land resources.

For many years, the soil-climatic classification did not become the basis of any policy-level intervention. The Planning Commission pioneered the much-needed shift in strategy at the end of 1980s, when it initiated a growth strategy based on the agro-climatic regional planning approach.

The purpose of this exercise was to identify broadly the crop mixes and types of land use that were best suited to the available resources of climate, soil, topography, water resources and irrigation facilities in each region. This implied a serious departure from the legacy of crop-centered research, which culminated in the Green Revolution. In the 1980s, Indian policy shifted from food self sufficiency to livelihood security. Acceleration of agriculture growth, with a special focus on improving the position of small farmers and extending the productivity revolution to non-irrigated areas was seen as a critical part of the strategy.

Area treated (million ha) and investment in Rs (crore)

Programme	Area treated till March 2005	Investment in Rs (crore) till March 2005
National Watershed Development Project for Rainfed Areas	7.95	2,398.76
River Valley Project and Flood Prone Regions	6.09	1,894.12
Watershed Development Projects in Shifting Cultivation Areas	0.28	236.35
Alkali Soil Reclamation	0.56	82.54
Externally Aided Project	2.36	4756.26
DPAP	15.13	2623.40
DDP	5.71	1857.78
Integrated Watershed Development Programme	6.32	2161.81
EAP	0.36	212.67
Integrated Afforestation and Eco-development Projects Schemes	0.82	813.73
Total	45.58	17,037

The country achieved considerable success with this approach in the 1980s.

Growth of agricultural gross domestic product (GDP) accelerated to about 4.7 per cent in the 1980s, compared with only 1.4 per cent in the 1970s.

This agricultural growth, together with the beginning of economic reforms in the non-agricultural sector, pushed up the growth rate of overall GDP to around 5.8 per cent in the period 1980-81 to 1989-90 compared to about 3 per cent in the 1970s. However, actual performance since the mid-1990s has been disappointing.

Agricultural growth slowed to 2 per cent a year in the 9th plan period, and overall economic growth was only 5.5 per cent, well below the 8 per cent target. During the 10th plan agriculture growth went below 2 per cent.

Programme	**Area treated till March 2005**	**Investment in Rs (crore) till March 2005**
National Watershed Development Project for Rainfed Areas	7.95	2,398.76
River Valley Project and Flood Prone Regions	6.09	1,894.12
Watershed Development Projects in Shifting Cultivation Areas	0.28	236.35
Alkali Soil Reclamation	0.56	82.54
Externally Aided Project	2.36	4756.26
DPAP	15.13	2623.40
DDP	5.71	1857.78
Integrated Watershed Development Programme	6.32	2161.81
EAP	0.36	212.67
Integrated Afforestation and Eco-development Projects Schemes	0.82	813.73
Total	45.58	17,037

WHY HAVE RAINFED REGIONS REMAINED OUT OF POLICY MAKERS' FOCUS?

The Green Revolution's Orphan

The biggest problem of rainfed areas is a historic policy mistake: adapting the Green Revolution principle for rainfed areas. It focused on intensive area approach comprising the already endowed areas. An anomaly at present is that large areas of land and large number of people in the country are outside the framework. "It is no wonder that these people also constitute the poverty geography of the country," says Planning Commission member B N Yugandhar.

In the last decade there has been a sharp deceleration in Indian agriculture with the growth rate of agriculture GDP slipping from 3.62 per cent from 1984-85 to 1995-96 to less than 2 per cent from 1995-96 to 2004-05. This is because irrigated areas, like Green Revolution states of Punjab and Haryana, have exhausted their potential.

"India's net sown area under crops has fallen drastically for the first time. The area under canal irrigation has fallen again for the first time in history. This is overall bad news for agriculture but worse for rainfed farmers," says former planning minister Y K Alagh.

Food grain production during the 10th Five-Year Plan was less than during the 9th Five-Year Plan period. From 1949-50 to 1964-65 (pre-Green Revolution period) additional lands were brought under the plough with extension of irrigation facilities.

From 1967-68 to 2002-03 (post-Green Revolution period), the scope for extension of cultivation gradually declined. The annual compound growth rate was quite low for all principal crops (0.29 per cent).

While India's population is increasing by 2 per cent every year, area under food production has shrunk by 12.5 per cent over the last decade. Thus food availability in rural areas has fallen to 152 kg per capita, 23 kg less than in the 1990s, creating large-scale food insecurity in villages.

The annual per capita production of cereals has declined from 192 kg in 1991/1995 to only 174 kg in 2004-07 and of pulses from 15 kg to 12 kg (1991/1995- 2004/07). This means that per capita food grain production is now at the 1970s level. This is because irrigated areas, like the Green Revolution states of Punjab and Haryana, have exhausted their potential.

On the other hand, the vast 60 per cent rainfed areas remain unexplored. According to the Planning Commission, the largest slump occurred in those areas that are predominantly rainfed.

Prime Minister Manmohan Singh didn't even hesitate to say that the slump in rainfed areas is a major policy blunder.

Why does the much talked about Green Revolution need to focus on rainfed areas?

What is the way Ahead?

Rainfed areas will have to be the focus of India's future agriculture revival. But as the past shows we need a different paradigm of development. "In view of their characteristics rainfed areas require approaches to agricultural development that differ from the Green Revolution strategy," says Katar Singh, chairman of the Anand-based India Natural Resources Economics and Management Foundation.

"There is an urgent need to change the mindset of the Green Revolution era, which focused more on inputs and did not look at regional variations. Because even water rich state like Kerala now get drought," says J S Samra, head of the newly-established National Rainfed Area Authority.

There is considerable potential to enhance productivity of rainfed areas. "Most of the areas will be able to take up a second crop," says Samra. Realising this potential essentially hinges on our ability to reverse the process of degradation — the process that will contribute to institutionalising water conservation, reducing run-off and erosion. Secondly, the crops must be suitable for local agro-climatic zones. Thirdly, market access to rainfed crops.

One of the major challenges facing rainfed agriculture in India today is to conserve and enhance the inherent capacity of its land and other natural resources to sustain agriculture. Any erosion of this inherent capacity will threaten India's food security and agriculture substantially. In order to constantly address this concern along with increasing production of food grains and other agricultural products, it is necessary to do the following: a) enhance and conserve the stock of available land, water and other natural resources, b) develop improved technologies, which maintain and improve the productive capacity of natural resources.

However, the first and foremost shift should be from 'input centricity' in the agriculture policy to 'needs/requirements' centricity. "By and large the prescription is the same as the good old Green Revolution — hybrid seeds, chemicals and mechanical inputs, monocrops. Much of the chemical inputs are a substitute for 'labour'. These lead to externalisation of inputs and can be devastating for rainfed farmers," says Satheesh of Deccan Development Society, a Medak-based NGO.

In its recommendations for improving the drought prone area programme, the Hanumantha Rao committee had in 1994 suggested that any area development programme must be location specific and must be based on the current status of natural resources such as vegetation cover and ground water recharge.

The decrease in rainfed areas or the increase in irrigated areas has been due to the increase in ground water use. But as the trend shows, recharge of ground water is hardly being addressed in the current watershed development programmes, the major initiatives to treat rainfed areas. Thus a greater focus has to be on recharging ground water in such areas along with surface storage and management.

For improving the water conservation strategy, there has been a suggestion of providing a labour component under National Rural Employment Guarantee Act (NREGA) for digging ponds in private lands. NREGA could be used for producing biomass even on private land for improving soil fertility.

"The government must provide water conservation subsidy to farmers in rainfed areas under NREGA," says Y S Ramakrishna of Central Research Institute for Dryland Agriculture.

"Time has come to integrate dryland food grains into the Public Distribution System and other food for work programmes so that dryland agriculture products get a market," says G V Ramanjaneyulu, of the Centre for Sustainable Agriculture, a Hyderabad-based NGO.

Livestock is an essential component of dryland ecosystems. Farmers get their livelihood during the drought mainly through livestock. Livestock, especially bullocks and buffaloes provide much needed farmyard manure for maintaining soil fertility apart from being useful for other agricultural operations such as ploughing and transportation.

Moreover, all livestock support systems are 'milk' centric. "There are practically no support systems available for livestock rearing for most farmers of dryland regions. It is time we thought about the role of livestock beyond milk," says J Venkateswarlu, former director of Jodpur-based Central Arid Zone Research Institute.

Even experts have been arguing about shifting a section of the population dependent on agriculture to non-farming sectors such as agro-processing, manufacturing and allied activities. "We need to create livelihood opportunities in agro-processing industries and manufacturing sectors," says Mahendra Deb, Director, Centre for Economic Social Studies, a Hyderabad-based research institute.

13

Policies Toward GM Crops in India

INDIA'S GM CROP OPPORTUNITY

The Indian people are far better fed on average than in the past, but 2.7 million children still die in India every year, 60 percent of them from diseases linked to malnutrition (Sharma 1999). A leading cause of malnutrition in India is poverty, and in rural areas a leading cause of poverty is low productivity in agriculture. Of the nation's one billion people, two thirds still gain their livelihood from farming, and seventy five percent of India's farmers are disadvantaged, with one hectare of land or less (Swaminathan 1999). Meanwhile continued population growth continues to increase the burden on Indian agriculture year after year.

The productivity of Indian agriculture improved in the 1990s compared to the 1980s, as the annual rate of growth of value added in farming increased from 3.1 percent to 3.8 percent, or roughly twice the nation's rate of growth of population (World Bank 2000). Poorly managed government marketing policies occasionally generated embarrassing public stocks of food, including 27 million tons of surplus wheat at the end of 2000. Yet the impression of national food abundance was misleading. During the decade of the 1990s, total food grain production in India did not increase at all on a per capita basis, and 230 million Indians remained food-insecure due to persistent poverty, linked most often to the low productivity of their agricultural resources.

Solving India's poverty and hunger problems will require more than just a boost in farm productivity, of course. A number of other issues will also have to be addressed, including persistent rural illiteracy, social marginalization, landlessness, and caste or gender discrimination. Even where low farm productivity is the problem, GM crop technologies might not be the solution. On India's drylands, where farm productivity is low in part due to poor soil fertility or scant rainfall, the GM technologies currently in use provide few new options to address such problems.

India's poorest farmers are those that live in dryland areas with less than 750 mm of rainfall a year who lack an ability to irrigate their crops. Non-irrigated farming in India still accounts for 67 percent of total cultivated area, and supports 40 percent of the population, plus two thirds of the nation's livestock. Average grain yields on non-irrigated land in India are only 0.7 to 0.8 tons per hectare, which is only one third the yield level on irrigated land (Singh and Venkateswarlu 1999). With yields on irrigated land now plateauing, India has little choice but to seek new technical solutions for its low-production farmers in dry, rainfed areas. In India 80 percent of millets and pulse crops are grown under rainfed conditions, and roughly 50 percent of rice is rainfed.

GM crops might seem an unlikely solution for farmers in hot drought-prone regions, since it has been far easier so far to engineer crops for specific resistances to pests or disease than to engineer the multi-gene traits needed to provide greater resistance to drought or heat. Yet India's producers of dryland crops (such as sorghum, groundnut, or pigeon pea) also face severe pest and disease problems, along with problems such as drought or heat. For groundnuts and pigeonpea, crop losses to biotic stress are actually greater than losses to abiotic stress (ICRISAT 1992). Pigeonpea farmers can sometimes lose their entire crop through damage from a single insect. Pod borers attack all pulses, and viral diseases are a widespread blight on India's dryland crops. Small dryland cotton farmers in India are devastated by bollworm infestations. Together with conventional breeding programs and improved training in integrated pest management (IPM), genetic engineering might help provide solutions to these biotic stress problems facing India's poorest farmers.

Environmental protection imperatives also argue for a GM crop revolution in India. The current practices of India's poor dryland crop farmers are damaging to rural ecosystems. If GM crops could produce yield gains for these farmers, there would be less need to clear new lands in rural India, plow fragile slopes, or destroy still more habitat. If farmers had insecticidal GM crops they also might escape having to risk their own health, pollute the environment, and kill so many non-target species as they do now with conventional chemical sprays. Farm chemical use is also a rural economic welfare issue in India, where cotton farmers currently spend 16 billion rupees annually on insecticide sprays. Vegetable producers in India currently suffer a $2.5 billion loss annually to insect damage, even while spending (on tomatoes, for example) $100-$200 per hectare on insecticides (Padmanabhan 2000)

Somewhat farther into the technological future, GM crops could eventually help address some of India's severe nutritional problems as well. Roughly 50,000 children in India go blind every year from vitamin A deficiency, while iron deficiency is a major threat to the health of women. The possibility of engineering iron-rich rice or vitamin A-rich rapeseed oil would become interesting in this context. Some traditional food crops in India that currently contain dangerous substances (neurotoxin in kesar dal, cyanide in tapioca, alfatoxins in groundnut) might also be rendered safer if genetic engineering could be used to "silence" these undesirable traits (Prakash 1999). And in a nation where refrigeration facilities are not yet abundant it might at some point be attractive to engineer fruits and vegetables less prone to spoilage. Tomato producers in India today lose 20-30 percent of their crop through postharvest spoilage.

Political leaders as well as scientists and technocrats in India have noticed these opportunities, and they now routinely endorse the potential contributions that biotechnology - including transgenic crops - might make to agricultural productivity growth and poverty reduction in the years ahead. In 1999, Dr. R. K. Pachauri, Director of India's Tata Energy Research Institute, stated that India's future increases in food production "would necessarily have to come from the application of biotechnology" (Pachauri 1999, p. 10). At India's 87th Science Congress in January 2000, Prime Minister Atal Bihari Vajpayee singled out information technology and biotechnology, plus "other knowledge-based sectors" as the propellers that would move India's economy ahead in the new century. At this same meeting Dr. R. A. Mashelkar, Director General of India's Council of Scientific and Industrial Research (CSIR) specifically endorsed biotechnology in agriculture as a means to turn farmers into high-productivity "knowledge workers" (BGE 2000).

While many top leaders in India have endorsed the value of agribiotechnology in general, and while scarce treasury resources have even been allocated to promote GM crop research within India's national agricultural research system, India's policies toward GM crops have hardly been promotional across the board, or even permissive. It was the original intent of biotechnology policy leaders in India to pursue an essentially permissive approach toward GM crops, yet this intent has recently been frustrated. Critics of GM crops were able to work within India's open and democratic political system to push for a precautious or even a preventive approach toward GM crops instead, especially in the area of biosafety policy. Indian biosafety authorities, somewhat like their counterparts in Brazil, ran up against forceful public criticism when they attempted to pursue a permissive approach toward the testing and release of GM crops. As of 2000 this meant that farmers in India, identical to their counterparts in Brazil and Kenya, had not yet been given official permission to plant any GM crops.

While GM crop critics in India have focused most heavily on possible biosafety risks, they have also attempted to block the technology by voicing strong objections to what they describe as the disadvantageous IPR implications of GM crops. To the extent that GM crops come with stronger IPR conditions attached, it is not surprising to hear voices of objection raised in India, a nation without a strong IPR policy tradition, least of all in the area of animals and plants. And to the extent that private multinational seed companies are the holders of so many of these IPRs, India's misgivings are that much easier to understand, given the culture of mistrust toward international companies that has long pervaded both India's government and its civil society. India's IPR policies toward GM crops, in part for these larger social and cultural reasons, have thus been extremely weak. As of 2000, India's Parliament still had not enacted even a basic plant variety protection law.

For decades India could get along without a strong IPR policy in the area of farm crops because it relied so exclusively on its own public sector scientists and government extension agents, rather than private investors or international private companies, to stimulate innovation in the area of agriculture. During the 1960s and 1970s, agricultural R&D in India was almost exclusively the business of the Indian Council of Agricultural Research (ICAR), a vast network of government institutes, directorates, research centers, and public universities. The private sector was effectively excluded through tight controls over germplasm exchange, technology imports, seed trade regulations, and restrictions on investment (Selvarajan, Joshi, and O'Toole 1999). In this environment, innovation was stimulated through public expenditure, rather than through the offering of IPRs to plant breeders. India operated without a plant variety protection law, and its 1970 Patent Act specifically excluded patenting of life forms such as plants (Mishra 1999).

This system served India well enough in the early years of the green revolution, and in some respects it continues to work well, as marginal economic rates of return to agricultural research in the state-dominated Indian system have remained high (Evenson, Pray, and Rosegrant 1999). By its own count during the 1992-96 Eighth Plan period, ICAR developed and released more than 2300 high yielding varieties and hybrids, including 452 new varieties of field crops ready for general cultivation. Yet by the 1980s, overall state support for agricultural research in India had begun faltering, and scientists in India began looking for ways to seek greater rewards through the private sector. Government attitudes toward the private sector also became less hostile. Starting in the early 1980s, the Government decided to provide its own breeder seed to private companies for multiplication and duplication. Foreign investment rules were relaxed, and in 1988 a New Seed Policy liberalized industrial licensing policies. As a result multinational seed companies began making significant investments, through India's own private sector companies. From the mid 1980s to the mid 1990s, 24 companies representing roughly 17 percent of all private seed concerns in India initiated technical or financial collaboration with foreign companies.

These liberalizations brought pressures in India to enact a basic plant variety protection law. Officials in the agricultural research establishment had concluded by 1991 it was necessary and prudent to move India's IPR policies closer to international standards. By 1993 a draft plant variety protection act (PVPA) was ready for submission to Parliament, one which measured up to the UPOV 1978 standard by creating relatively weak PBRs and by preserving the farmer's privilege to replant and exchange saved seed. The drafting of this basic PVPA preceded final negotiation of the TRIPS agreement in WTO, but when India formally ratified TRIPS in 1994 these issues became linked, since India now had an international obligation to have basic IPR protections in place for plant varieties by January 2000.

This Government decision to move toward a conventional plant variety protection law in the context of TRIPS triggered a surprisingly emotional debate in India's Parliament. The first draft of the PVPA was criticized by the private seed industry for being too weak, but NGOs claiming to represent farmers groups warned it was far too strong and would lead to a private expropriation of farmers' rights. Revised drafts were produced in 1996-97 in order to address the farmers rights issue, and Cabinet approval for a revised draft was secured in October 1997, but under still more NGO criticism Parliament continued to stall, and still more redrafting was initiated. The version of the PVPA that was working its way through Parliament in 2000 was a version produced in December 1999.

These PVPA debates in India would have been contentious even if GM crop technologies had never been invented. In India, the experience of colonization was an especially bitter one and sensitivities toward foreign ownership of any part of the national economy - especially by foreign companies - remain high. Foreign corporate ownership of genetic resources in India - GM or otherwise - is an especially emotional issue. Advocates of the PVPA have argued that IPR guarantees for plant breeders will stimulate useful new innovations in India, bring in foreign capital and technology, and at the same time provide for improved protection for the 150,000 accessions and samples of crop germplasm in the hands of India's own National Gene Bank. Existing plant varieties that are already in the "public domain" will also not be affected by the law. Yet critics highlight what they describe as the danger of a foreign corporate takeover of India's own national seed industry, or the private use of IPR options to gain controlling ownership of the plant genetic resources on which India's farmers depend. An Indian NGO calling itself the Gene Campaign argues prominently that foreign multinational companies would be able to appropriate and exploit India's genetic material without adequate compensation if western-style IPRs in this area were ever to be permitted (Sahai 1999).

Fueling these fears after 1991 was the increased presence of international companies staking ownership claims within India's food economy. Between 1987 and 1995, even though a PVPA had not yet been passed by Parliament, the share of private seed sales in India made by firms with foreign ownership increased from 10 percent up to 33 percent (Pray et al 1991; Pray and Kelley 1998). Several IPR actions taken outside of India were also represented as threats to national genetic sovereignty. In 1992 the W.R. Grace company secured a patent in the United States for a distinctive chemical formulation of a naturally occurring pesticide from neem trees. Anti-corporate activists in India began asserting that international companies such as Grace were planning to use such patent protections to appropriate for themselves the locally developed folk-knowledge that had long been available to India's farmers and rural communities for free. Grace argued that this charge was unfounded, since the patent was on a new chemical formulation implying that it did not prevent traditional farmers all over the world from continuing to use neem extract as they always had. Nonetheless the neem case enflamed popular anxieties regarding the foreign appropriation of local knowledge, and led to an international struggle by environmental NGOs to remove the patent. In May 2000, the European Patent Office did revoke Grace's 1995 patent in Europe for the neem oil extraction process, but litigation against the patent in the United States was sure to prove more difficult.[1]

Some of India's most respected leaders in the area of agricultural research have shared the concern that patent protections for plants, or even a national move toward a conventional PBR system, might leave the nation's poor farmers at a disadvantage. Rural communities in India have for thousands of years employed their own on-farm seed selection practices to breed a highly diverse stock of plant varieties nicely attuned to local conditions. Under a PBR system, why should IPR protection go only to the professional breeders (working either within international companies or national institutes) who routinely use these already improved local varieties as the basis for their breeding programs? Arguing that indigenous knowledge systems are similar to general scientific information, in that they are a valuable part of public knowledge, Dr. M.S. Swaminathan, India's most acclaimed agricultural scientist and the first winner of the World Food Prize, has helped popularize the notion that communities of farmers are as entitled to IPR protections for their efforts as professional breeders. Largely in response to his leadership, and as early as 1989, the FAO Conference in Rome adopted his concept of Farmers Rights.

Swaminathan also saw to it that India's own draft PVPA law would incorporate a farmers' rights provision, in the form of a institutional mechanism designed to ensure the rights of farming communities to financial compensation for any earlier contributions they might have made to the plant varieties that will finally be protected under the act. The 1999 draft of India's

PVPA (in Chapters X and XII) thus certifies the rights of villages and local communities to claim contributions to the evolution of plant varieties, and to then receive appropriate financial compensations paid from a centrally managed National Gene Fund (Government of India 1999). In anticipation of this provision coming into effect, the Government of India began compiling a formal register of the geographic origins of all plant genetic resources that have been or might be used by professional breeders. Registering this indigenous germplasm also puts it in the public domain, thus preempting private bioprospectors from subsequently using the PVPA to take commercial advantage of this germplasm without compensating local communities.

As innovative and ambitious as these provisions for financial compensation to farmers might be, they have not gone far enough to satisfy India's most determined critics of UPOV-style plant variety protection. In 1998, India's activist Research Foundation for Science, Technology and Ecology (RFSTE) headed by Vandana Shiva, who is an internationally known opponent of GM crops, the green revolution, and all other market-led scientific advances in Indian agriculture. RFSTE proposed an alternative Biodiversity Rights and Protection Bill that would give local farming communities not just financial compensation but actual property rights over the crop varieties to which their traditional on-farm breeding efforts had made contributions. The Bill specifies that local communities would share these property rights with the Central Government of India so that private companies suspected of seeking to pirate away valuable genetic materials would not be able to approach local communities separately to gain access to these materials on unequal terms (Cullet 1999). The intent of this proposal was to reject any movement in India toward a western-style IPR system in the area of plant genetic resources; when the Government persisted with its PVPA approach, RFSTE initiated a public interest litigation action (PIL) against the Government.

The issue of GM crops only adds new complication to this already intense internal debate in India over plant variety IPRs. In 1998, when the Monsanto Company of the United

States purchased a 26 percent share of India's own Maharashtra Hybrid Seeds Company Limited (Mahyco), a Monsanto executive was quoted in the Indian press saying "We propose to penetrate the Indian agriculture sector in a big way. Mahyco is a good vehicle." (*The Economic Times*, New Delhi, April 26, 1998). Opponents of transnational corporations within India took this as a direct challenge, and began directing harsh criticism at all of Monsanto's GM crop technologies, especially the "terminator gene" patent it had recently acquired, which was presented as a direct threat to the tradition of seed saving in India. It was Monsanto's misfortune that word of this new gene use restriction technology (GURT) reached India just at the time limited trials of Mahyco's Bt cotton were being authorized for the first time. A Canadian based NGO, the Rural Advancement Foundation International, or RAFI, spread the alarm that Monsanto's terminator gene might be turned loose in India. Local NGOs opposed to international companies and non-traditional farm technologies were thus mobilized easily against GM crops. In November 1998, a local political leader in Karnataka state who had previously gained attention by attacking both the Cargill seed company and a Kentucky Fried Chicken restaurant in Bangalore, staged for the media a brief attack on some of Mahyco's Bt cotton field trials. Soon thereafter local communist party leaders in Andra Pradesh pressured the Chief Minister into calling for an end to the Mahyco Bt cotton field trials underway there as well.

Gene use restrictions technologies had not yet been inserted into GM crops anywhere let alone the Bt cotton being tried out in India, but this issue made Monsanto and Mahyco easy targets for NGO and opposition party criticism. The mere existence of a terminator gene patent seemed to confirm suspicions that international seed companies were seeking to take away from India's farmers their traditional right to replicate seed on their own farms. In India, 92 percent of all wheat seed planted is home grown, and 88 percent of paddy rice seed. NGOs and globalization critics feared that India's small farmers would be pressured by Monsanto or Mahyco into purchasing expensive GURT seeds, only to discover too late that they had to keep purchasing them year after year. The same argument could have been made against conventional non-GM hybrid seeds, which are currently used and repurchased annually by small as well as large farmers in India growing maize, sorghum (jowar), millet (bajra), sunflower, cotton, and vegetables, but the terminator technology was far more inviting as a proxy for criticizing profit-making foreign companies. Monsanto announced in the fall of 1999 that it was not going to commercialize the technology, yet by then the damage was already done, including damage to the Government's continuing efforts to move the PVPA law through Parliament.

Enactment of the PVPA in India was also made more difficult by links to the TRIPs agreement in WTO. Many of the same groups who saw IPRs for plants as the opening wedge for external corporate domination of India also feared the WTO, which is often criticized in India as an instrument used by rich countries to bully poor countries. Under the TRIPS agreement India was technically obliged to change its patenting system from processes to products, and to have a plant variety protection law in place by January 2000, but this deadline came and went without parliamentary action.

Prospects for eventual Parliamentary passage of some version - probably a weakened version - of the Government's PVPA bill were nonetheless strong in 2000. The National Democratic Alliance government sent its most recent version of the bill to Parliament for approval in December 1999, just before its January 2000 WTO deadline for TRIPs compliance. It was then referred to a thirty-member joint committee of both Houses which redrafted the bill so as to strengthen its farmers' rights provisions, by recommending that appeals be heard not by the High Courts but by a specially formed tribunal. The delays continued, but WTO was unlikely to criticize India for not having a PVPA in place so long as the delay seemed to reflect the slow movement of democratic procedures.

This failure of India through 2000 to set in place any formal IPR protections for plants was not, however, the principal reason that GM crop technologies were not yet in use by farmers. Because of the size of India's commercial seed market, private international companies have at times been eager to bring GM technologies into the country even in the absence of IPR guarantees. Their preferred means for doing this has been to seek to introduce hybrid varieties, which carry their own inherent biological protections against seed saving and replanting.

It has not so far been a corporate unwillingness to bring valuable proprietary GM crops into India that has kept this technology out of farmers' hands. India's weak IPR policies do, however, place some limits on what international private companies are willing to do, as well as on what India's own breeders have an incentive to do. Because of weak IPRs, private companies do not like to conduct advanced research in India labs. In its Bt cotton venture the Monsanto company conducted no plant transformations within India itself; instead it brought in hybrid cotton seeds that had already been transformed abroad. Nor did not begin field trials until it owned a secure share of a local corporate partner that had been entrusted to do the backcrossing. A Belgian biotechnology firm, Plant Genetic Systems, was almost as cautious when it decided to bring its transgenic "seedlink" technology into India, to develop an improved *brassica* (mustard). It invested an initial US$1 million in a joint venture with a trustworthy local private sector partner, Pro Agro, and even so it limited its transfer of transgenic materials to local Indian hybrids only.

If India wishes to partner with foreign companies to secure access to GM technologies for plants other than hybrids (for example, glyphosate resistant soybeans), passage of a credible plant variety protection law will be critical. Up to a point, bilateral contracts can be used to compensate for a weak internal IPR law, yet without at least a UPOV 1978 plant breeders rights system in place, India's public sector researchers risk being passed over as unsafe technology development partners by the more dynamic international private sector. Guarding India's own nationally managed germplasm accessions from private appropriation could also become more difficult, public registration strategies notwithstanding. For these reasons India's Council for Scientific and Industrial Research (CSIR) has been campaigning for a stronger IPR and patenting system in India, in agriculture as well as in other sectors.

In the longer run, IPR guarantees even stronger than those contained in the current draft PVPA may be necessary if India wishes to keep pace in GM science. The draft PVPA has been criticized by the private seed companies as unattractive for their purposes in several respects. The draft PVPA lays a basis (in Chapter X) for compulsory licensing after three years to meet "the reasonable requirements of the public for seeds." It creates a registration process open to delay and challenge which could add another three years to the already long time period (currently 6-7) years required in India to get a new seed variety completely de-regulated and eligible for commercial release. And by embracing the UPOV 1978 standard the PVPA does not provide scientists with sufficient incentive to innovate at the molecular level, since protection of GM varieties may not be extend to included essential derivations from those varieties. The President of India's private Seed Association, while a supporter of the PVPA, has stated his view that it "does not provide sufficient protection for bioengineered plants." (Selvarajan, Joshi, and O'Toole 1999, p. v). Private seed companies would of course prefer product patents as the best means to encourage innovations in biotechnology, but Indian scientists working at the molecular level inside the national research system also want stronger standards; they would prefer the UPOV 1991 standard over the weaker UPOV 1978 standard contained in the current draft PVPA .

Without a plant variety protection law formally in place, India as of 2000 must be classified as taking a preventive stance toward GM crops in the IPR area, and even after the draft PVPA is eventually enacted India's posture in this area will be no more than precautionary. Still, it has been India's biosafety policies, not its IPR policies, that have most conspicuously slowed down the nation's GM crop revolution.

BIOSAFETY POLICY

Biological safety has not been, historically, a strong public policy thrust in Indian agriculture. Rural biological systems have long been threatened by low resource farmers who clear new lands and destroy natural habitat in order to graze animals or plant crops. Mismanagement of irrigation water has left once fertile croplands poisoned with salt. Careless application of chemical nutrients has contaminated rivers, streams and ponds, making water unsafe for human consumption and poisoning fish and amphibian species. Improper use of insecticides has killed non-target species while making target pests increasingly resistant to the poisons. Even after the enactment of India's powerful 1986 Environmental Protection Act (EPA), such conventional biohazards from farming were seldom put under effective regulation. It is thus somewhat surprising to see Indian officials paying so much attention to the biological safety of GM crops. India's official interest in genetic resource IPRs would have been strong even without GM crops as an issue, but the nation's current official interest in rural biological safety is to a large extent a product of the GM crop revolution itself.

India's first embrace of a formal national biosafety policy dates from the first emergence of rDNA agricultural and human health technologies in the 1980s. We might expect, given India's strong development aspirations and its prior history of low attention to the rural biohazards, that the nation's biosafety policies toward GM crops would have been permissive or even promotional, rather than precautionary or preventive. Instead, the Indian Government adopted from the start a set of procedures that ensured significant precaution, and as implemented - in the context of a growing global controversy over GM crops - these procedures sufficed to slow the progress of the technology considerably. As of 2000 farmers in India were not yet permitted to grow any GM crops, because none had yet received final approval from the nation's biosafety regulators.

The Indian Government first issued rules and procedures for handling GM organisms in December 1989. The Department of Biotechnology, inside the Ministry of Science and Technology, published these rules and procedures in January 1990 (Department of Biotechnology 1990). These "Recombinant DNA

Safety Guidelines" were revised slightly and republished as "Revised Guidelines for Safety in Biotechnology" in 1994 (DBT 1994). They describe the biosafety measures that must be undertaken in India both for contained research activities and also for large scale open environmental release of genetically altered agricultural and pharmaceutical materials. A subsequent 1998 revision of the Guidelines elaborated procedures for screening transgenic plants and seeds for toxicity and allergenicity (DBT 1998).

India's crop biosafety Guidelines were written to require a screening of GM crop technologies for scientifically demonstrated risks, borrowing heavily from the demonstrated risk approach employed by USDA's Animal and Plant Health Inspection Service (APHIS) in the United States. This has given the Guidelines considerable permissive content on paper. Yet the implementation of the Guidelines is ultimately shared with India's Ministry of Environment, a procedure which inclines India's policy somewhat more toward precaution. The Guidelines create two separate review committees: a Review Committee on Genetic Manipulation (RCGM) which is empowered to approve (or not approve) applications for all *small scale research activities* in India designed to generate information on transgenic organisms, and a Genetic Engineering Approval Committee (GEAC) empowered to approve (or not) *large scale research activities, plus actual industrial use or environmental release* of all GM organisms.

Because the RCGM mostly screens research activities for biosafety, it is composed of representatives from India's leading public universities and research Institutes, including the Indian Council of Agricultural Research (ICAR), the Indian Council of Medical Research (ICMR), the Council of Scientific & Industrial Research (CSIR), and others. It is constituted by the Department of Biotechnology (DBT), and its member-secretary is adviser to DBT. It has thirteen members in all, and operates by majority vote. In contrast the GEAC is a statutory body under the Ministry of Environment and Forests (MoEF) and is empowered to approve or disapprove all large-scale use and environmental release of GM organisms. GEAC is thus India's most powerful biosafety policy gatekeeper. It is chaired by the Additional Secretary of MoEF, co-chaired by an expert nominee from DBT, and it includes representatives from DBT, from the Ministry of Industrial Development, the Ministry of Science and Technology, and the Department of Ocean Development. GEAC is designed to perform more than a technical function. It is the lead interministerial body empowered to shape - by consensus - the Government's final disposition toward large-scale use and environmental release of GM organisms. The GEAC is empowered to authorize or prohibit, conditionally or unconditionally, the import, export, transport, manufacture, processing, use, or sale of any GM organism (Ghosh and Ramanaiah 2000).

This Indian biosafety system was intended to be permissive at the research stage, where RCGM was given the final word. Since RCGM is constituted by DBT - the agency explicitly tasked with promoting biotechnology in India - its membership comes from the very research institutes (in some cases the very individuals) who conduct biotech research and who rely on DBT for funding. When RCGM hands the regulatory process over to GEAC, after the research stage, a precautionary bias takes over, since GEAC is chaired by MoEF. This built-in tension between permissive research versus precautionary release might have some advantages in theory. By allowing research to go forward, India's process is capable of generating the necessary empirical data (e.g., from field trials) needed to make sound final release decisions(Ghosh 1997). India's system may also be potentially useful for avoiding last minute jurisdictional disputes between biotechnology commissions and environmental ministries, of the kind seen in Brazil between CTNBio and IBAMA. In India, biosafety regulation stays legally under the EPA and within the jurisdiction of GEAC from start to finish. The public sector dominated Indian system also excludes representatives of private industry from direct participation, another contrast to Brazil, and it requires that RCGM send a Monitoring Committee to visit the experimental sites that are being employed by applicants to generate biosafety data (DBT 1998).

India's Guidelines are quite thorough regarding the biosafety questions that must be answered prior to commercial release of GM technologies. In the area of transgenic plants, basic information must be provided on the characteristics of the donor organisms, the vectors used, the transgenic inserts, and then on the resulting transgenic plants themselves. Through lab, growth chamber, greenhouse and field trials, evaluations must be generated by applicants regarding toxicity and pathogenicity, possibility and extent of transgenic pollen escape and transfer to wild relatives, and consequences for both the environment and human and animal health. The data may be generated in India or anywhere else (Ghosh 1997). Guidance is provided on proper (biosafe) procedures for conducting these experiments, along with the proper format for presenting this information as a formal registration document. Some of the data required by India's procedures go well beyond biosafety. RCGM is required in the latest version of the Guidelines to solicit from applicants and assemble information "on the comparative agronomic advantages of the transgenic plants" to assure GEAC that the technologies under review are "economically viable" as well as environmentally safe (Department of Biotechnology 1998, p. 6). While this might seem an added barrier to approval, it could also favor approvals in some cases by allowing evidence of social benefits to be presented alongside evidence of possible environmental risks.

This being the structure of India's biosafety review system for GM organisms, how has it operated in practice? In the area of health care, a number of GM products have been reviewed successfully with little controversy. As of March 2000, a total of ten different transgenic health care products in India (two locally produced, 8 imported) had moved all the way through the system and had been cleared for biosafety deregulation by GEAC. Social confidence regarding the biosafety of GM health care products has been consistently high in India. In the area of transgenic plants, however, the system has not moving so smoothly. Under criticism from NGOs, even RCGM has been forced into a highly cautious posture, and as of 2000 GEAC had yet to grant a commercial release for any GM crops.

India's biosafety review system for GM plants was pushed into this cautious posture in 1998, when RCGM was accused by NGOs in India of having exceeded its mandate by giving approval to limited field trials for Mahyco/Monsanto's Bt cotton plants. The original Guidelines had been ambiguous as to where RCGM's authority to manage small-scale research activities would end and where GEAC's authority to manage large-scale activities and "environmental release" would begin. When Mahyco applied to conduct contained field trials of its Bt cotton in 1998, RCGM assumed it had the authority to approve such trials, in part because it had earlier approved contained field trials of transgenic mustard in 1995 without incident. Yet the Bt cotton field trials approved for 40 locations in nine different states India in 1998-99 soon became a focus of intense national controversy, as noted above. NGOs in India, led by RFSTE, claimed that only GEAC had authority to approve field trials of GM crops, since even contained and limited trials constituted an environmental release. In February 1999, RFSTE then filed a PIL against the DBT for the manner in which it had authorized the trials (and also against Mahyco plus several other ministries), asking - without success - that further field trials be blocked.

RFSTE's challenge to the Bt cotton field trials was motivated not so much by any specific biosafety worry. The "Bollgard" Bt cotton seeds that Monsanto had brought in to cross with local Indian varieties had previously undergone extensive testing and monitoring in the United States to meet the biosafety standards set by APHIS and EPA. Bt cotton had been grown successfully in the United States on a commercial scale since 1996, and by 1998 it covered 40 percent of total U.S. cotton area. By 1998 it was also being grown widely and successfully in Australia, South Africa, and China. Bt cotton was potentially attractive in India as a solution to the bollworm infestations that farmers could no longer control with chemical sprays due to resistance in the pest population. This emergence of a real pest resistance problem from non-GM technologies made it harder to invoke hypothetical pest resistance problems as a reason not to plant GM cotton. resistant pest population in India in response to non-GM cotton production techniques made it harder to invoke hypothetical "superbugs" as a reason not to use GM cotton. Food safety risks were also hard to invoke for Bt cotton, since cotton is an industrial crop rather than a food. RFSTE was using biosafety rules and procedures as its means to try to block Bt cotton, but the core of its argument against this GM crop was that it was being introduced by the Monsanto company from abroad, the company which owned rights to a dreaded "terminator" technology.

Yet RFSTE also sought to make its case by pointing to procedural irregularities in RCGM's approval of the field trials. RCGM was faulted for having given field trial authority directly to a private company applicant, and for then allowing that applicant to launch the trials directly on leased farmers' fields, rather than on government research farms under direct public sector management. While this would seem a logical way to handle an application filed by a private company, RCGM might have known it would be seen in India as trusting the private sector too much. Second, RCGM had not secured advance approval from government authorities in several of the states where the trials were located, and in at least one of those states the trials even began before a local biosafety committee had been fully constituted. These several irregularities were exploited by RFSTE, and left RCGM and DBT politically isolated when fears regarding the terminator gene subsequently broke. Nor did it help that RCGM's procedures in this case had been largely non-transparent. RCGM could point out that there was no formal obligation to publish deliberations or invite public comment prior to approving the limited trials, but when the trials became controversial RCGM and DBT were criticized all the more heavily for having operated behind closed doors (Raj 1999). In the midst of this controversy in 1999, DBT's biosafety Guidelines were amended to clarify that GEAC's approval would be required for all field trials larger than one acre per location, or larger than 20 acres per year nationally. DBT also directed that the 1999 field trials for Mahyco's Bt cotton would take place either on university research farms or under closer national institute supervision. This, plus a DBT-led public awareness campaign in 1999, plus ministerial promises that the terminator gene would not be allowed into India, helped calm the public mood considerably.

Although India's Bt cotton trials went forward under protest and criticism, they did manage to generate important evidence of the GM plant's effectiveness against bollworm infestations. These pest attacks on cotton in India had emerged as a visible social issue due to the growing resistance of the pests to chemicals, and the resulting desperation of some small farmers. Excessive spraying of diluted and adulterated chemical insecticides - in some parts of India as many as 14 sprayings a season - had brought on the pest resistance problem, and some poorer cotton farmers in India who had borrowed heavily at a high rate of interest to buy chemicals that no longer worked were falling heavily into debt. Their plight came to public attention when bad weather worsened the pest crisis in 1998, and as many as 500 farmers resorted to suicide as their only way out (Sharma 2000). The first 1998 field trials of Bt cotton in India had seemed to confirm, meantime, that Bt varieties could provide at least a short-term solution to the pest control problem. DBT reported in 1998 that on average (in controlled field trials planted in eight different states in India) Bt cotton was able to reduce insect damage dramatically: yields for Bt cotton were 40 percent higher than for the non-Bt controls, and with an average of five *fewer* chemical sprays (Ghosh and Ramanaiah 2000; James 2000). For an Indian farmer with 5 ha of cotton, such a reduction in chemical use from switching to Bt cotton might by itself represent a savings of roughly Rs 2000 per crop (about $50).

RCGM was pleased and reassured by the 1998 field trial data, but nonetheless asked Mahyco in 1999 for ten more field trials, seeking more data on commercial advantages to farmers and also on a hypothetical geneflow concern, the possibility that insects could take Bt cotton pollen far enough away from the plant to result in unwanted gene transfer to other plants. Only after these additional trials had been completed did RCGM, in April 2000, formally express its technical confidence in Bt cotton. RCGM noted that its latest evidence showed farmers growing Bt cotton obtained 25-75 percent higher yields, while using six fewer chemical sprays, and with no evidence of harm to crops in adjacent plots.

Having received this approval at the research stage from RCGM, Mahyco promptly applied to GEAC for permission to begin large scale trials, and in response to this application, GEAC did approve large scale field trials for Bt cotton (up to 85 total hectares) in July 2000. Additional acreage was also permitted for seed production, in anticipation of a possible commercial release as early as 2001. To appease protesters GEAC made large scale field trials conditional on obtaining an independent Indian laboratory certification that the cotton plants did not contain the "terminator" gene. Even so, RFSTE was antagonized and submitted an additional petition against the trials. GEAC's prompt field trial approval increases somewhat the likelihood of an eventual commercial release for Bt cotton in India, although the first release may be only within a tight acreage restriction, once again on grounds of biosafety caution.

Has the Indian system been precautionary in this case only because the Bt cotton is a Monsanto variety? A second application working its way through India's system, from another company (ProAgro-PGS) to develop transgenic hybrid mustard, also met a cautious response from RCGM. Contained field trials of GM mustard began in 15 different locations in 1995, followed by open field trials, and none of these field trials was ever disrupted by an anti-GM activist, yet the process of bioscreening was highly precautionary just the same.

In 1999, India's regulators asked ProAgro-PGS for one more year of field trials to screen for effects on soil micronutrients, an issue that had never been highlighted before. ProAgro-PGS hoped for eventual deregulation from GEAC, but recognized that it may be a conditional release only, perhaps with a requirement (hard to enforce) that farmers planting the GM mustard seeds separate themselves at least 40-50 meters from non-GM fields, as an extra precaution against unwanted geneflow. While India's anti-GM NGOs may be focused almost exclusively on keeping Monsanto's products out of the country, biosafety regulators in India have thus been highly precautionary toward all GM crop applications, whether from Monsanto or not.

TRADE POLICY

Since GM crops are politically controversial in India, the government is under pressure from NGOs to impose a GM-free trade policy requirement on the nation's commodity imports and exports. So long as no GM crops are being planted within India on biosafety grounds, such an official GM-free trade policy can be relatively easy for the government to embrace and implement. A policy of standing aloof from international GM commodity trade is also relatively easy to embrace in India because the nation has, for decades, tended to stand somewhat aloof from all international food commodity trade in its official pursuit of "national food self-sufficiency." Ever since India's bad experience with excessive dependence on international food aid in the 1960s, political leaders have sought to avoid not only renewed concessional dependence on world markets, but commercial dependence as well.

India's policy aversion to international food trade of all kinds is reflected in the fact that the nation has recently accounted for roughly 10 percent of total world agricultural production, but less than 1 percent of world commodity trade (Sharma 2000). With continued income growth in India, demand for imported grains is likely to grow, yet actual imports will continue to be slowed by significant trade barriers at the border. As one example, India recently imposed an 80 percent duty on rice to curb the influx of what it called "cheap grain."

India occasionally imports small quantities of corn, but it strictly controls these imports with a tariff rate quota imposing a 60 percent duty on imports larger than the quota. In the case of wheat, India allows imports only in rare cases to offset specific internal transport cost problems, for example to allow some less expensive imported wheat to reach some coastal flour mills in the southern part of the country. India also exports very little wheat, despite its occasionally large internal surplus stocks. In 2000, when India's excess wheat stocks reached 27 million tons, efforts were finally made to clear the stocks through export, yet these were frustrated in part by the known presence of a wheat crop disease - "Karnal Bunt" fungus - in some parts of India, which has put wheat from India on the import ban list of some thirty other countries (APBN 2000a). Given these larger aversions in India to free international food commodity trade, and given the absence so far of any GM crop production within India itself, imposing an effective ban on the import and export of GM commodities has been easy enough for the Indian government to manage. Under India's official GM Guidelines it is GEAC which must approve any large scale import of GM commodities, and so long as GEAC has not approved the commercial planting of GM crops at home it may naturally wish to go slow on approval of commercial imports from abroad. Politicians in India will resist GM crop imports as well. Ever since the 1998 controversy over field trials for Monsanto's Bt cotton, government ministers have routinely defended themselves with statements that GM foods are not being grown *or imported* into India, and will not be imported without proper safeguards.

India's official GM crop import ban was tested under unusual circumstances in the summer of 1998, when the nation experienced a highly publicized food safety emergency unrelated to GM crops. More than 50 people in Delhi died after eating a contaminated locally grown (and non-GM) mustard oil, the most popular cooking oil in northern India. The government acted by banning sales of mustard oil pending safer packaging. To make up the shortage, it had to authorize temporary imports of soybeans and lower its import duties on soybean oil (*Wall Street Journal, December 8, 1998, p. A10).* Anti-trade and anti-GM NGO leaders responded to this action with a letter to the Prime Minister, alerting him to the fact that some of the soybeans about to be imported could be GM. The government sought to finesse the issue on this occasion by arranging to import soybeans that were already split and hence suited only for oil production, and not planting. Because the soybeans were not intended for environmental release and since they were not explicitly labeled GM, the Government on this occasion managed to avoid seeking formal import approval from GEAC. In the future it may be more difficult for the Government of India to avoid engaging GEAC in approval of bulk commodity imports. The January 2000 Biosafety Protocol obliges exporters (under Article 18) when shipping LMOs internationally to label those shipments as "may contain LMOs" and "not intended for intentional introduction into the environment." Identifications of this kind are more likely to trigger a mandatory review in India by GEAC, or at least more vigorous complaints by NGOs if GEAC is again avoided. NGOs in India opposed to GM crops and foods can take extreme positions on the import question. In June 2000, RFSTE reacted with outrage when it learned that some of the corn-soya blend food aid that was being imported via CARE and CRS to provide relief to Indian victims of a super cyclone in the Eastern coastal state of Orissa, had come from the United States and was thus likely to be "genetically contaminated." RFSTE castigated the United States for "using the Orissa victims as guinea pigs for GE products" and called on the government of India to explore alternative sources of food aid (RFSTE 2000).

India's policy regarding imports of GM germplasm for research purposes has so far been permissive rather than preventive. Some extra steps are required when importing GM materials for research, and some bureaucratic delays are encountered, but the imports themselves have never been held back. This partly reflects the fact that RCGM, rather than GEAC, has final authority to clear transgenic imports for research purposes (DBT 1998, p. 8). Technically, any importer of GM germplasm must also get phyto-sanitary clearance from India's relevant quarantine authority, the National Bureau of Plant Genetic Resources (NBPGR), yet NBPGR typically acts on the recommendation of RCGM and conducts a phyto-sanitary evaluation, as a routine precaution against import of diseased or infested materials.[5] RCGM is a committee dominated by representatives from the government institutes doing some of the research, so its permissive stance on imports is not surprising, yet private companies as well have had little trouble bringing GM materials into the country if the purpose is research only.

The import decision process in India is prone to a jurisdictional dispute between RCGM and GEAC, similar to that discussed earlier in the case of field trials. RCGM is authorized to approve small scale imports only, and GEAC alone can approve large scale imports, but there is not yet a clear definition of the dividing line between the two. Tiny quantities of 10 kg or less have until now fallen clearly under RCGM's jurisdiction, but a potential for conflict over larger shipments remains. India's Guidelines document empowers GEAC to give approval or disapproval (from an environmental vantage point) of all large scale "import, export, transport, manufacture, process, [or] selling of any microorganisms or genetically engineered substances or cells including food stuffs..." (DBT 1990 p. 17) The encompassing language of this provision again suggests how little the Government of India has been willing to entrust this new technology to private sector market forces alone, let alone international market forces.

In export markets, India has on occasion faced a temptation to use its GM-free status to seek price premiums. India is a small exporter of soybean meal (1.5-2.2 million tons per year, in recent years), and it has recently promoted these meal exports (also sunflower meal and rapeseed meal) as "GM free" for sale to overseas markets, such as Japan, Indonesia, Thailand, the Philippines, the Gulf countries, and the Middle East (APBN 2000b). Yet most of these sales are for animal feed purposes, so price premiums have been difficult to secure. Indian meal exporters have nonetheless begun hoping that Asian countries such as Thailand, because they export chickens to the GM-conscious European market, will soon see the advantage of taking feed imports from a GM-free supplier such as India, rather than from nations that plant GM crops such as the United States.

India can thus be classified as embracing a trade policy toward GM crops which is fully "preventive" by the definition in use here: a de facto blockage on GM commodity imports, coupled with occasional efforts to use the nation's GM free status to seek premiums in export markets. The strength of this policy will be tested if and when GM crops are finally given biosafety approval for planting by Indian farmers, particularly food crops (not just cotton). At this point the nation's overall GM free status will disappear, so costly internal market segregation and labeling procedures will be required in order to offer GM free guarantees to foreign customers. Avoiding these costly trade-linked procedures could become an additional reason for India's GEAC to go slow on the final approval of internationally traded GM food crops for planting by India's farmers.

FOOD SAFETY AND CONSUMER CHOICE

As of 2000, India had not yet developed an explicit consumer food safety policy toward GM crops. This partly reflected India's status as a low-income developing country where many poor consumers were still concerned as much about the availability or the price of food as about its exact characteristics or ingredients. Yet this also reflected India's official claim that GM foods were not yet being sold in the country. In this regard, India was in a similar position to Kenya. If the Government of India begins approving GM food crops (not just cotton) for commercial production, it may need to develop a more explicit food safety and consumer policy as an accompaniment.

Consumer protection laws have been slow to emerge overall in India, partly reflecting the nation's poverty yet also partly reflecting a large role played by the state itself in providing consumer goods through state owned industries. The state has historically been reluctant to sponsor a strong consumer protection movement partly for fear of liabilities to its own public industries. The notion of protecting consumers by giving them an "informed choice" has also been slow to develop in India, partly because India's citizens and its government officials have historically been comfortable with a more direct regulatory approach. Consumer advocates in India tend to focus more on issues of state regulatory failure rather than informed choice. In the food safety area, for example, the state is routinely held accountable by opposition parties or through the media for occasional failures to prevent the contamination or adulteration of processed and packaged foods. As noted above, a 1998 dropsy outbreak in Delhi linked to contamination of edible oils prompted a quick tightening of packaging rules. Partly because so much food consumption in India has traditionally been satisfied from unpackaged natural ingredients, consumer fears regarding adulteration of packaged foods remain understandably widespread. According to one survey, 60 percent of Delhi's population fears that branded milk could be adulterated, and hence unsafe to drink (Rao 1999).

Because India does not yet officially grow or import any GM foods, it has been able to continue operating within food safety policies that draw little or no distinction between GM and non-GM food ingredients. India's still operative 1954 Prevention of Food Adulteration Act predates the GM crop revolution and does not mention transgenic entities. In 1998, however, India made at least one adjustment to the GM crop revolution when it revised its biosafety approval guidelines to require that GM seeds, plants and plant parts be separately screened for toxicity and allergenicity (DBT 1998). By introducing a separate regulatory procedure for GM crops, this new RCGM guideline gives India a permissive rather than a fully promotional safety policy toward GM foods. The protocols published in the revised guidelines do not set a higher food safety standard because of the GM nature of the products; they were developed by RCGM through consultation with conventional industrial toxicologists and scarcely go beyond the testing that might be appropriate for judging the safety of pesticide residues.[6] Also, the new procedures allow for test results generated in other countries to be used in India, on the grounds that food safety testing does not have to be site specific. This means the data used by FDA to give approval to a GM food in the United States could presumably be re-used for the same purpose in India. The process thus can be a quick one, and RCGM officials have already indicated informally that in the GM mustard application (the first food crop application going through India's approval system) their review of the information submitted has already found no new food safety risks (Ghosh and Ramanaiah 2000).

Labeling policies in India have been a moot question until now because of the nation's nominal GM-free status and because most food consumption continues to be satisfied through home or street preparations of natural foods that are never packaged at all, let alone carefully labeled. Foods grown in India for export purposes must be labeled according to the policies of the importing countries, yet the nation's GM-free status has transformed this requirement into something of a competitive advantage for the country rather than a burden. Without any costly market segregation, all of India's soy or castor oilseed cake destined for Europe and Japan can be labeled "GM free."

Labeling requirements are likely to be imposed as soon as GM food crops (as opposed to fiber crops like cotton) are released by GEAC for commercial planting by India's own farmers. Transgenic pharmaceutical products in India that have been deregulated for sale be GEAC already must be labeled GM. Anticipating the GM labeling problem for foods, India's Ministry of Food Processing Industries has approached DBT seeking advice, and India's Ministry of Health and Family Welfare has approached both DBT and the Codex Alimentarius in Rome asking how the nation's 1954 Prevention of Food Adulteration Act can be adjusted to take GM ingredients into account. If and when GM foods do officially come onto the market in India, DBT can be expected to advocate labeling requirements that are permissive rather than precautionary, to avoid a need for full market segregation.

PUBLIC INVESTMENTS IN RESEARCH

The Government of India, principally through its Department of Biotechnology, has for more than a decade directed a small but steady stream of treasury resources toward the development of its own GM crop varieties, as well as toward the simple backcrossing of foreign-developed GM traits into local germplasm. Between 1989-97, DBT spent a total of nearly 270 million rupees from the treasury (roughly $6 million) on plant and molecular biology research, with projects focused primarily on development of transgenic plants (Ghosh 1999). India's public research investment policies toward GM crops thus deserve to be classified here as "promotional." Tangible scientific payoffs from these investments have been slow to develop, however, reflecting both the modest total size of the investment plus some longstanding limitations within India's public sector research establishment. Without greater public spending on research, and without significant institutional reforms and policy adjustments, including some to help public researchers partner more effectively with the domestic and international private sector, India's goal of developing its own commercially useful GM crop technologies could remain elusive.

As noted earlier, India has a long and distinguished history of making effective public investments in agricultural research through the national institutes and universities that operate under ICAR. Public research by state level governments in India has also grown since the mid-1970s, even to rival total ICAR spending at one point in the mid 1980s. India's public sector investments in agricultural research, together with its formidable extension system, have had large impacts on productivity growth. By some estimates these impacts from research and extension have been larger even than those from government investment in rural infrastructure, education, irrigation, or power (Fan, Hazell, and Thorat 1999). Agricultural research and extension investments have been the source of roughly three fourths of all agricultural productivity growth in India in recent decades (Evenson, Pray, and Rosegrant 1999). Not all agricultural research in India has been public sector-led. After the mid-1980s, India's more liberal international investment policies helped attract more private agricultural R&D spending into the country. Between 1987 and 1995, private seed company R&D expenditures in the country increased from 41.7 million rupees to 154.9 million rupees. Still, total public sector agricultural R&D in India continued to exceed private sector R&D by a wide margin of roughly 6 to 1 (Pray et al 1991 and Kelley 1998).

Modern biotechnology was first introduced into the Indian research community in the 1970s, when a significant number of Indian scientists trained in the United States and Europe returned to take up work in India (Balasubramanian 2000). Significant government investments began quite early as well, with establishment in 1982 of a National Biotechnology Board. In 1986 this board was established formally as India's Department of Biotechnology inside the Ministry of Science and Technology (Sharma 1999). A significant part of DBT's mission has always been to distribute and finance specific research contracts, and in the area of agriculture DBT set up six centers for plant molecular biology in different regions of the country and has provided research funding to those centers-primarily through ICAR - for the improvement of specific priority crops such as rice, mustard, chickpea, pigeonpea, and wheat (DBT 1999).

DBT must secure its budget every year from the Planning Commission and the Ministry of Finance, and the resources it receives are quite modest, despite the fact that senior political leaders frequently list biotechnology as among the keys to India's future economic growth and prosperity. In 1998-99, the total research budget of DBT across all agricultural and non-agricultural areas was 1,040 million rupees, or roughly US$26 million. About 15 percent of this total (US$3.8 million) went for all plant biotechnology, including medicinal and aromatic plants, tree and woody species, and pilot plants for tissue culture. DBT's investments in *transgenic* plant biotechnology in 1998-99 totaled only about US$ 1.3 million.

With these modest treasury investments, DBT has been attempting to fund a significant spread of GM crop research activities, including not only basic transformation work but also research in extremely challenging areas such as abiotic stress (e.g., drought tolerance) and improved nutritional quality (e.g., protein quality, oil quantity and quality, or higher starch content). For example, ICAR is making efforts at Jawaharlal Nehru University (JNU) in New Delhi to engineer potato and rice with improved protein content, and elsewhere it is using GM to modify the starch content of potato and the Vitamin E activity of mustard and groundnut (Paroda 1999). A number of more standard GM applications are also being researched, including Bt varieties of rice, cotton, pigeonpea, potato, and mustard. National researchers have also worked with transformed eggplant, cabbage, cauliflower, and tomato.

As ambitious as this GM research agenda is, most of the work so far is at an early stage, or confined to laboratory or greenhouse trials only (Ghosh 1999, Table 1). Of all the ICAR GM varieties under development, DBT as of 2000 had approved field trials only for GM eggplant and mustard. The Indian Agricultural Research Institute (IARI) and the Central Institute for Cotton Research (CICR) have produced their own transformed Bt cotton, but as of 2000 just 100 plants had been grown in the laboratory only, and several years of backcrossing with local varieties would have to take place before any full scale field trials.

Over time, there is little doubt that India's national researchers will be able to develop and field test their own GM crop varieties suitable and ready for farmers to use. Prior to commercial release, however, some significant IPR impediments are likely to be encountered. Many of the transgenes India's researchers have been inserting are owned by foreign companies and have been made available for research purposes only. In the case of India's GM rice, the transgenes come from IRRI free of charge, but IRRI in turn has them for research purposes only via MTAs from the private patent holders. When the time comes to consider a commercial release in India, negotiations will be necessary with these international patent holders. ICAR presently has no clear strategy to break through such looming IPR bottlenecks, and it has yet to undertake the kind of IPR literacy program for its researchers that Embrapa earlier undertook in Brazil.

In several other respects ICAR may face difficulties pursuing its independent course toward GM crop technology development. ICAR continues to be dominated by conventional plant breeders, many of whom were trained years ago, well before the latest advances in GM technologies (Murthyunjaya and Ranjitha 1998). Younger scientists have been trained to do GM work for ICAR through the Rockefeller Foundation's rice biotechnology program, but often these people are hired away by the private sector, which pays more than twice the public sector salary. Younger scientists with good ideas can get support from DBT via ICAR, but excessive paperwork and long delays are routine. DBT's budget is not only small; it also tends to be spread too thin since each separate institute in the national system lays its claim to at least some financing. For example, India's rice transformation budget is divided up between at least five different locations within the national system, which weakens the impact at any one location.

India's national system takes pride in its independence yet some of its best work in the area of agribiotechnology has resulted from international donor support. Beginning in 1988 the Rockefeller Foundation became an important catalyst, helping India's national institutes and universities to build infrastructure, train younger scientists, yet this important source of external support has recently been phased out. The United States Agency for International Development (USAID) is also providing assistance, since December 2000 through a new joint research project run by Michigan State University and India's Tata Energy Research Institute (TERI) in Delhi, working to enhance the beta-carotene content of mustard oil to fight blindness caused by Vitamin A deficiencies. Also, the Ministry of Foreign Affairs of the Netherlands has funded an Andhra Pradesh-Netherlands Biotechnology Programme for Dryland Agriculture since 1996, engaging several ICAR institutes in a fruitful partnership to bring biotechnology applications to non-irrigated low-resource farmers producing sorghum, castor, groundnut, and pigeon pea.

This project, supported at 18 crores Rs over its first six years, has been more successful than the parallel biotechnology programme funded by the Dutch in Kenya, and discussed earlier in Kenya, so it is now scheduled to be extended for another 10-15 years. The programme focuses on low-end village level biotechnology applications, such as biopesticides and biofertilizers, but it has also given some financial support to ICAR researchers working on insecticidal GM varieties of sorghum and castor (IPE 2000). ICAR also has benefited from a substantial National Agricultural Technology Project (NATP) loan from the World Bank to upgrade its scientific skills, infrastructure and management procedures, and to facilitate better links between the component parts of the fragmented system, but little of this money seems to be reaching the labs of younger researchers in the biotechnology area. Meanwhile, links between ICAR and the private sector remain difficult to establish (Hall, et. al 1998).

Partnerships with the international private sector should be an attractive option in India, where national researchers could offer international companies not only their rapidly improving transformation skills and their long established conventional breeding skills, but also access to Indian germplasm from the 150,000 accessions and samples in India's National Gene Bank, established by ICAR as a part of India's National

Bureau of Plant Genetic Resources. When private companies show interest in working with ICAR or DBT, the response is muted however. The Monsanto Company at one point offered to share its Bt technologies with ICAR free of charge for transformation and production of orphan crops such as chickpea and pigeon pea within India, but ICAR showed little interest. And just as ICAR is leery of partnering with the private sector, most international companies are leery of depending too heavily upon ICAR, which has a reputation for slow movement of scientific results from the lab to the market place. The international private sector has shown its preference to work in India not with ICAR but with local joint venture partners in the private sector. In 2000 when Monsanto agreed to share its GM technology for high beta-carotene mustard with Indian scientists, it agreed to work through the private Tata Energy Research Institute, not through ICAR. Private seed companies in India have not only been ahead of ICAR in commercializing improved varieties such as hybrid rice; they are also known to have control over the highest quality hybrid germplasm. The international private sector will also be shy of working with ICAR on GM crops so long as the other policies of the Government of India - especially in the areas of IPRs, biosafety, and trade - remain highly precautionary or preventive rather than permissive. This is a loss, since the chances of getting a new GM crop variety through India's recently cautious biosafety screening process would probably be higher if the application for deregulation came via ICAR, rather than from a purely private applicant such as Mahyco/Monsanto.

Summarizing, the desire of the Government of India to promote GM crop technologies primarily through public sector research investments is in some respects logical and commendable, since private sector companies are unlikely to make the GM crop research investments needed to enhance nutritional traits for the malnourished, or to address abiotic stress production problems encountered in semi-arid regions by low resource farmers. Yet the success of this strategy will remain in doubt so long as the treasury resources being invested are so small, so long as the public sector institutions in question retain their current limitations, and so long as the policies of the government in other critical venues - biosafety in particular-remain preventive or highly precautionary.

CONCLUSION

We can see from this summary a seeming contradiction between India's public research policy toward GM crops, which is promotional, versus its IPR, biosafety, and trade policies, which are precautionary or even preventive. Is there an explanation for this apparently inconsistent mix of policies? Given India's unmet farm productivity needs, and given the endorsement of GM crops by its own research community, why the extreme precaution in these other areas? India's sluggishness in developing a plant variety protection policy is perhaps understandable because it links to so many issues other than GM crops, and because it has not in any case been the highest barrier to getting GM crops into the country. But how can we explain India's strong precaution toward GM crops in the two areas of biosafety and trade policy?

A comparison to some of India's policies at the outset of the green revolution in the middle years of the 1960s can perhaps shed some light here. In 1965, when the Center for Improvement of Maize and Wheat (CIMMYT) in Mexico and the International Rice Research Institute (IRRI) in the Philippines first made high yielding varieties of wheat and rice available to India, a political debate ensued within the country not so different from the current debate over GM crops. Indian groups on the political left and those claiming to speak for the interests of small farmers were highly skeptical then toward the green revolution HYVs, just as these same groups today tend to be highly skeptical toward GM crops. The new Mexican wheat varieties had not yet been widely tested in India itself, so critics argued it might be dangerous to begin planting them on a large scale without more information regarding their local performance and impact. The new varieties also called for more sophisticated management practices that small farmers might not be able to master, and a use of purchased inputs small farmers might not be able to afford, so fears arose that only big commercial farmers would gain, and India's rural poor might be made poorer still. The new varieties also required more fertilizer than India itself could produce at the time, so critics feared an implied dependence on fertilizer sales made by private international agrochemical companies located in Europe or the United States. Despite these many understandable and sincerely voiced concerns, the Government of India set caution aside in 1965. Under courageous political leadership from Agriculture Minister C. Subramaniam and others, India decided in 1965-66 to import 200 tons of Mexican wheat and appropriate quantities of high yielding rice to launch an ambitious seed production and farm demonstration campaign, with a goal of planting 13 million hectares to the new varieties within five years. Subramaniam was told by western experts and by India's own Planning Commission that his goal was too ambitious, and that he should not try to move so fast (Subramaniam 1979, pp. 45-51). Subramaniam's plan to move ahead quickly with new green revolution seeds was ultimately supported by Prime Minister Indira Gandhi and then aggressively implemented, despite continued internal opposition. His approach won the policy debate on this occasion because the alternative to taking the new seeds was continued dependence on imported food aid from the United States, which had always come with unwelcome diplomatic strings attached. As Subramaniam recalls telling his critics at the time, "Would you like to continue this dependence on America for imported foodgrains? Was that preferable to raising domestic production? Instead of importing food was it not better to import fertilizer and plant protection chemicals to help raise production?" (Subramaniam 1979, p. 28).

When GM crop technologies first became commercially available in the 1990s, India was no longer significantly dependent on food imports from abroad, either as food aid or even as commercial purchases. The success of the green revolution itself had eliminated the need for such imports. GM crop advocates have therefore not been able to use Subramaniam's trump argument: that an embrace of GM crops is necessary to free India from food import dependence.

Instead of importing wheat as food aid, India's government today is burdened by surplus stocks of wheat. Several hundred million Indian villagers are still poor and poorly fed because they cannot improve the productivity of their own farmlands or protect their crops from pests and disease, and this could well be seen as a legitimate reason to move ahead quickly with GM crops. But this line of argument is hard to develop since most of India's leading advocates for the poor within the NGO community are opposed to GM crops on other, just as many of them were earlier opposed on similar grounds to the green revolution HYVs. RFSTE's Vandana Shiva, the leading NGO critic of GM crops in India today, had earlier made an international career out of criticizing high yielding non-GM crops in India (Shiva 1991). It was perhaps to be expected given these political dynamics, that a far more cautious approach to the new technology would prevail. In the current case the caution is expressing itself in India through controversies over the biosafety of GM crops, but the same larger background debate about the wisdom of relying on the international private sector or of importing new agricultural technologies from the West is alive and influential just below the surface.

But what is the opinion of India's farmers toward GM crops? Until farmers in India are given permission by the government to grown these crops, this important question will be impossible to answer. Farmers in India developed a strong positive opinion toward high yielding seed varieties in the 1960s and 1970s once they were given access to the new seeds, and once India's own farmers voted in favor of high yielding varieties by adopting the new seeds to quickly and so widely, the policy debate continued among some NGOs and non-farmers but otherwise became largely irrelevant. Today in India various self-styled farm leaders have taken pro-GM or anti-GM positions in the policy debate, but until GM crops are released for commercial use by actual farm communities the views of farmers will not be known. By slowing the movement of GM crops into farmers' fields, the government is postponing the day when the nation's most important stakeholders in the transgenic crop revolution will be able to develop and express their own informed opinion.

Bibliography

Adelberg, Edward A.: *Papers on Bacterial Genetics*, Boston, Brown and Company, 1960.

Ausubel, F.M.: *Current Protocols in Molecular Biology*, New York, John Wiley and Sons, 1989.

Bains, William: *Biotechnology: From A to Z*, New York, Oxford University Press, 1998.

Barnes, N.: *Biology*, New York, Worth Publishers, 1989.

Barnum, Susan R.: *Biotechnology: An Introduction*, Belmont, Thomson/Brooks/Cole, 2005.

Bauer MW: *Biotechnology-the Making of a Global Controversy*, Cambridge, Cambridge University Press, 2002.

Brandwein, P.F.: *Sourcebook for the Biological Sciences*, San Diego, Harcourt Brace JOvanovich, 1986.

Broach, J.R.: *The Molecular Biology of the Yeast*, Cold Spring Harbor, Cold Spring Harbor Laboratory, 1981.

Brooks, Rodney A.: *Flesh and Machines: How Robots Will Change Us*, New York, Pantheon Books, 2002.

Chrispeels, Maarten : *Plants, Genes and Crop Biotechnology*, Sudbury MA, Jones and Barlett Publishers, 2003.

Clark, J.M.: *Experimental Biochemistry*, New York, W.H. Freeman and Company, 1977.

David Sadava: *Plants, Genes and Crop Biotechnology*, Sudbury MA, Jones and Barlett Publishers, 2003.

Dixon, Dougal: *After Man-A Zoology of the Future*, New York, St. Martin's Press, 1981.

Dodds, John H.: *Plant Genetic Engineering*, New York, Cambridge University Press, 1985.

Drexler, K. Eric: *Engines of Creation: The Coming Era of Nanotechnology*, New York, Anchor Books, 1986.

Fransman M, Junne G, Roobeek A: *The Biotechnology Revolution?*, Oxford, Blackwell, 1995.

Friedberg, E.C.: *DNA Repair*, New York, WH Freeman and Company, 1985.

Fumento, Michael: *Bioevolution: How Biotechnology is Changing Our World*, San Francisco, Encounter Books, 2003.

Gaskell G: *Biotechnology-the Making of a Global Controversy*, Cambridge, Cambridge University Press, 2002.

Goodsell, David S.: *Bionanotechnology: Lessons From Nature*, Hoboken, Wiley-Liss, 2004.

Kurzweil, Ray: *The Age of Spiritual Machines*, New York, Penguin Books, 1999.

Moravec, Hans: *Mind Children: The Future of Robot and Human Intelligence*, Cambridge, Harvard University Press, 1988.

Morholt, E. : *Sourcebook for the Biological Sciences*, San Diego, Harcourt Brace JOvanovich, 1986.

Old, R.W. : *Principles of Gene Manipulation*, London, Blackwell Scientific Publications, 1989.

Poindexter, J.S.: *Strategies and Mechanisms for Field Research in Environmental Bioremediation*, American Academy of Microbiology, Washington, D.C., 1994.

Primrose, S.B.: *Principles of Gene Manipulation*, London, Blackwell Scientific Publications, 1989.

Rifkin, Jeremy: *The Biotech Century*, New York, Penguin Putnam, 1998

Shetty, Kalidas: *Food Biotechnology*, New York, Dekker/CRC Press, 2005.

Slater, R.J.: *Experiments in Molecular Biology*, Clifton, Humana Press, 1986.

Smith, John E.: *Biotechnology*, Cambridge, Cambridge University Press, 2004.

Switzer, R.L.: *Experimental Biochemistry*, New York, W.H. Freeman and Company, 1977.

Walden, Richard: *Genetic Transformation in Plants*, England, Open University Press, 1988.

INDEX

A

Achievements, 115, 156, 195.

Agency, 5, 21, 55, 85, 121, 149, 156, 176, 181, 182, 203, 256, 272.

Agricultural Mechanisation, 174, 175.

Agricultural Performance, 1, 2, 94.

Agricultural Watershed, 56.

Approach, 32, 40, 41, 42, 51, 62, 64, 70, 80, 90, 92, 101, 112, 113, 123, 134, 140, 143, 147, 153, 160, 161, 164, 168, 173, 180, 182, 194, 200, 202, 222, 224, 238, 239, 240, 247, 252, 258, 269, 276.

Approaches, 19, 41, 92, 97, 136, 190, 230, 238.

C

Coconut Development Programme, 115.

Communication, 75, 77, 78, 79, 82, 98, 100, 135, 191, 192, 214, 217.

Construction, 21, 79, 85, 150, 151, 161, 162, 168, 169, 196, 200, 204.

Cotton Development Scheme, 116.

Crop Opportunity, 241.

Crop Prospects, 95.

D

Developments, 93, 234, 281.

Diagnostic Analysis, 5, 6, 10, 11, 13, 16, 20, 38, 46, 47, 54.

E

Evaluation, 9, 10, 20, 21, 22, 23, 24, 40, 42, 46, 47, 49, 155, 163, 169, 189, 219, 264.

F

Food Processing Industries, 198, 203, 268.

Food Safety, 259, 263, 266, 267.

G

Guidance, 55, 204, 205, 257.

H

Horticulture, 4, 91, 93, 98, 100, 125, 134, 139, 143, 144, 145, 146, 147, 148, 149, 150, 153, 156, 157, 158, 176, 178, 184, 191, 202, 211, 281.

I

Information, 5, 6, 9, 10, 13, 14, 16, 17, 18, 19, 20, 21, 22, 24, 25, 26, 29, 32, 34, 37, 40, 41, 46,

47, 48, 56, 57, 63, 64, 75, 76, 77, 78, 79, 80, 81, 82, 87, 97, 100, 102, 134, 135, 141, 154, 155, 159, 160, 164, 165, 171, 172, 192, 198, 199, 201, 202, 204, 205, 214, 217, 244, 248, 255, 257, 267, 274.
Investments, 88, 89, 90, 91, 92, 213, 215, 218, 231, 234, 246, 268, 269, 270, 273.
Irrigation System, 5, 6, 7, 9, 10, 19, 37, 39, 41, 46, 131, 154, 165, 172, 175.

L

Land Degradation, 29, 30, 219.

M

Market Complex, 200, 201.
Mass Media, 98, 191.
Mechanisation, 174, 175, 176.
Mechanisms, 92, 228.
Methodology, 26, 47, 55, 62, 71, 97, 163, 169.
Micro Irrigation, 101, 154, 155, 166, 173, 175.
Mobile Mapping System, 75, 76, 77, 78, 80, 81, 82, 281.

O

Opportunity, 21, 93, 241.
Organisations, 160, 177, 214, 220.

P

Plant Protection, 85, 111, 142, 219, 275.
Plantation, 4, 125, 143, 147, 178, 202, 211.
Press, 130, 131, 132, 250.
Promotion, 100, 101, 123, 125, 127, 130, 131, 138, 147, 157, 158, 165, 172.
Public Investments, 88, 268, 269.

R

Rainfed Farming Development, 83, 84, 281.
Research, 4, 19, 28, 29, 30, 75, 84, 86, 88, 90, 121, 130, 131, 144, 145, 164, 171, 183, 184, 185, 186, 188, 189, 190, 204, 212, 213, 214, 216, 217, 218, 219, 220, 221, 224, 227, 229, 232, 233, 234, 239, 240, 244, 245, 246, 248, 249, 252, 253, 255, 256, 258, 259, 261, 264, 268, 269, 270, 271, 272, 273, 274, 281.

S

Soil Classification, 66, 109.
Soil Management, 109.
System Performance, 5, 7, 9, 20, 27, 37, 38, 40, 42, 45, 46, 47, 54.

T

Trade Policy, 262.
Trade policy, 262, 265, 274.

W

Water Management, 5, 6, 8, 12, 13, 14, 18, 19, 28, 29, 35, 39, 47, 52, 54, 151, 158, 159, 160, 218, 222.
Water Planning, 55.

□□□